Engaging Minds:
Rethinking Assignments for Critical Thinking in the Age of AI

by

Dr. Jeffery Adams, Ed.D, M.S., NRP

For permission requests, contact the author at:
jefferyad@icloud.com

Engaging Minds: Rethinking Assignments for Critical Thinking in the Age of AI
First Edition
Printed in the United States of America

ISBN: 978-1-7348825-8-2

Dedication

I dedicate this book to my wife, Alisha—
My partner in crime, in chaos, in service, and in every step forward.

We've built a life defined by grit and purpose, side-by-side in ambulances, simulation labs, and all the moments between. Through every late night, hard decision, and leap of faith, you've stood with me—not behind me, but beside me—pushing when I needed it, grounding me with honesty, and always keeping sight of the bigger picture.

You didn't give up on your dreams to support mine—you adapted, redirected, and made space for both of us to grow. Your passion, your sacrifice, and your resilience are woven into the story behind this book. *Engaging Minds* may have my name on the cover, but its roots are ours.

My success has always been your success, too.

Acknowledgments

Writing *Engaging Minds* has been a journey of reflection, creativity, and collaboration. This book would not exist without the encouragement, insight, and support of those who believed in both the message and the mission.

To my wife, Alisha—thank you for your unwavering support, your patience through long hours of writing, and your belief in what I'm building. You've always been my foundation and my motivation.

To Dr. David Kashmer—thank you for introducing me to AI and showing me its educational potential long before it was mainstream. That early spark led to years of exploration, design, and application that ultimately shaped the heart of this book.

To Dr. Philip Bryant—thank you for encouraging me to write this book. Your words were the push I needed to take scattered ideas and turn them into something coherent, useful, and lasting.

To the many colleagues, peers, and thought partners who took time to help me think through and review the ideas in this book—thank you. Your feedback helped sharpen the message and ensure the content reflects both vision and practicality.

To my students—you are the reason I keep doing this. Watching those "aha" moments, seeing you grow in confidence and clarity, and knowing that we're building more than knowledge—we're building thinking—means everything.

To my professional community in simulation and EMS, and to the faculty I've had the honor of working alongside at PCOM, VCOM, Duquesne, and now Lipscomb University—thank you for embracing innovation and for believing that education should always move forward.

And finally, to every reader who picks up this book looking for something more than a template—thank you. I hope the ideas here spark your own. Sometimes all it takes is one new perspective to change how we teach—and how students learn.

Prologue: The Why Behind This Book

—I've never thought outside the box—because I've never seen the box.

Growing up with undiagnosed dyslexia, learning was never a straight line. While others read through chapters once and moved on, I had to read every section multiple times, constantly breaking down content just to understand it. Over time, I stopped seeing learning as passive consumption and started treating every assignment like a problem to solve. I didn't realize it at the time, but this way of thinking would shape my entire career.

When I entered the world of medical simulation, something clicked. Simulation demanded that same kind of thinking—designing backwards, predicting what students would see, ask, or miss, and creating scenarios that *led* them to the right conclusions without giving them away. I didn't teach answers. I built experiences that forced students to ask better questions.

For over twenty years, I've worked in simulation across medical schools and institutions. I've trained faculty, built curriculum, developed national certification scenarios, and designed experiences to replicate the chaos of real clinical encounters. And in every case, the best learning happened when students had to think—not just click.

—So when AI showed up, I didn't see a threat. I saw a tool that finally worked at the speed of my ideas.

A colleague, Dr. David Kashmer, introduced me to GPT just a few weeks after it launched. I dove in, and within days, I was creating custom AI tools to streamline scenario writing, automate curriculum mapping, and even make manikins speak in riddles that students had to solve. What used to take hours of design time could now happen in seconds—with *more* creativity, not less.

But the more I engaged with other educators, the more I noticed that experts on AI in education were speaking generically, offering surface-level advice rooted in a pedagogical mindset that didn't fit today's learners. These talks weren't wrong—but they were outdated. Today's students are digital natives. They've had technology in their hands since birth. They know how to search. They know how to swipe. But they often don't know how to think critically about what they're seeing.

—As educators, we need to stop blaming the tools and start rethinking how we teach.

AI is not the enemy. The enemy is stagnation. We're still designing assignments for a world that doesn't exist anymore. Today's learners don't need another lecture—they need challenge, relevance, and clarity. They need assignments that demand original thinking, contextual application, and personal insight. And most of all, they need educators who understand how to build learning environments that leverage AI rather than fear it.

—That's why I wrote this book.

This is not a book of theories. It's a book of examples. You'll find practical strategies, real-world frameworks, and assignment designs that work in the AI era. But more than that, I hope this book shifts the conversation—from fear to possibility. From generic to specific. From passive to engaged.

—Because if we want students to think, we need to design experiences that demand it.

Table of Contents

Chapter 1: Introduction

The first signs of AI's influence in medical education did not emerge through formal instruction—they appeared in admissions. Faculty and staff began approaching me with a shared concern: they suspected applicants were using AI to write their personal statements, but they lacked a clear way to prove it. The writing felt off—too polished, too vague—but they were unsure how much weight to assign that instinct or how to assess the authenticity of what they were reading. They needed a way to distinguish genuine experiences from AI-generated content, and they asked for help. Drawing on my experience in simulation—where observing behavioral nuance is central to evaluating performance—I recommended a simple change. We started requiring applicants to include specific clinical experiences in their statements. That adjustment made a difference immediately. During interviews, the divide became obvious. Students who wrote from experience spoke with clarity. Those who relied on automation stumbled. The solution wasn't punitive—it was diagnostic. It didn't ban AI. It exposed gaps in understanding. And like simulation, it taught us that well-designed prompts can do more than filter—they can force depth. Across my work designing scenarios, training faculty, and building curriculum, I've learned that effective education is never about catching students—it's about constructing moments where shortcuts fail and thought becomes necessary.

1.1 A Paradigm Shift: Moving Beyond Fear of AI

Discussions about artificial intelligence (AI) in education often emphasize potential risks, including academic dishonesty, student dependency, and ethical concerns. Although legitimate, these risks frequently overshadow the opportunities AI presents to enhance educational experiences. Institutional responses have historically focused on restrictive measures, such as banning AI tools or relying on academic integrity pledges. However, these approaches typically address symptoms rather than root causes. Research suggests that student disengagement, rather than technological misuse, often drives academic integrity issues, indicating a need for more thoughtful strategies (Nguyen & Brown, 2018).

Reframing AI as a resource rather than a challenge can yield transformative outcomes in education. By integrating AI into assignment design, educators can create tasks that require deeper engagement and critical thinking. For instance, broad prompts such as

"Analyze the economic effects of renewable energy adoption" enable AI to generate plausible yet generic responses. In contrast, a prompt like "Assess the impact of renewable energy policies in your region using data from local government reports" demands unique inputs and analysis that AI cannot easily replicate. This shift in design reduces the risks of misuse and encourages authentic student engagement (Anderson & Miller, 2021).

Evidence shows that incorporating AI into learning can boost student participation and creativity. Assignments involving structured reflection, contextual analysis, and real-world applications foster meaningful learning experiences. For example, tasks requiring students to critique AI-generated outputs or compare them to human-created solutions can highlight AI's limitations while reinforcing critical thinking. Studies suggest that such approaches discourage academic dishonesty and promote active, ethical engagement in learning (Greenfield, 2022).

Embracing AI's potential aligns with the broader goal of preparing students for an AI-driven world. Research highlights AI's ability to enhance personalized learning, streamline administrative tasks, and provide immediate feedback—allowing educators to focus on higher-order instructional strategies. By leveraging these strengths, educators can transform classrooms into dynamic spaces where AI becomes a tool for exploration and innovation rather than disruption (Thomas, 2021).

Addressing AI's educational challenges and opportunities requires a balanced approach that integrates ethical considerations, robust assignment design, and student-centered learning strategies. As education evolves, so must our frameworks for understanding and responsibly applying AI. By fostering environments that encourage critical thinking and deep engagement, educators can transform potential risks into opportunities for meaningful learning. This approach ensures students are prepared to thrive in a future where AI is integral to both personal and professional domains (Allen & Rivera, 2023).

1.2 A New Perspective: AI as an Ally

Artificial intelligence (AI) is often viewed with skepticism in education due to concerns about misuse, such as academic dishonesty and diminished critical thinking. While understandable, these apprehensions often stem from an incomplete understanding of AI's role. Shifting the narrative to see AI as a supportive tool can unlock its potential to enhance learning and innovation. Research suggests that thoughtful AI integration promotes active engagement and critical thinking rather than bypassing intellectual effort 【3】【6】.

A primary concern is that AI enables shortcuts, allowing students to bypass meaningful engagement. However, studies show that misuse often results from poorly designed assignments rather than from the tool itself. For example, a prompt like "Describe the importance of teamwork" invites generic, surface-level responses easily generated by AI. In contrast, asking students to "Reflect on a group project you completed during this course, describing specific communication challenges and how your team resolved them" requires personal experience and course-specific insight—elements AI cannot fabricate convincingly 【6】【7】.

Constructive AI integration can support higher-order thinking and improve student engagement with course content. Educators can design assignments that involve critiquing AI-generated responses or comparing them with human-authored work, encouraging students to assess accuracy, tone, and relevance. These tasks teach students how to evaluate the usefulness of AI and recognize its limitations—skills increasingly relevant in professional settings. Using AI as a tool for brainstorming can also help students generate initial ideas, but it should be accompanied by reflective discussion. For example, after a brainstorming session that involved AI, instructors can prompt students to consider whether the energy use behind the technology justifies its benefit. Questions such as, "Was using AI for this task worth the environmental cost?" or "Did this experience remind you of how effective peer brainstorming can be?" shift the focus from convenience to thoughtful use, helping students make informed decisions about when and why to engage with these tools 【3】【10】.

Viewing AI as an ally benefit not only students but also educators. It offers opportunities for personalized feedback, adaptive learning, and streamlined administrative tasks, enabling educators to focus on creative and strategic aspects of instruction. Evidence shows that thoughtful AI integration can boost engagement, spark creativity, and improve student satisfaction 【11】【12】. By fostering an environment where AI complements human intellect, educators can better prepare students to navigate the complexities of an AI-integrated world.

Reframing AI from a source of apprehension to a tool for innovation is essential for modern education. Assignments and learning environments shaped by this perspective emphasize active participation, critical engagement, and responsible technology use. This approach ensures students develop the skills needed to integrate AI effectively into their personal and professional lives, empowering them to succeed in an AI-driven future【6】【9】【12】.

1.3 Moving Beyond Integrity Statements

Traditional methods for maintaining academic integrity often rely on honor codes, integrity statements, and punitive measures. While these strategies underscore the importance of ethical behavior, they fall short in addressing the unique challenges posed by AI in education. Given AI's accessibility and capabilities, institutions must adopt proactive approaches that go beyond rule enforcement. Rethinking assessment design, teaching methodologies, and AI integration of AI into the educational process.is essential for fostering genuine academic integrity.【3】【6】

Integrity statements alone rarely engage students in understanding why academic honesty matters. Research shows that simply reiterating rules has little impact on behavior unless paired with meaningful learning experiences that highlight the value of original thought and intellectual effort【7】. For instance, students asked to analyze AI-generated outputs compared to their ideas are more likely to recognize AI's limitations and appreciate the importance of unique insights. This approach promotes academic honesty and encourages deeper engagement with the material【8】.

Moving beyond integrity statements requires designing assessments that promote active learning and critical engagement. Tasks involving collaborative problem-solving, real-world applications, and personalized inputs reduce opportunities for academic dishonesty while making learning more meaningful. For example, an assignment like "Create a project proposal based on the community needs identified in your fieldwork" compels students to draw on their experiences and observations, minimizing the utility of AI for shortcuts. Research consistently emphasizes the role of well-designed assessments in fostering intrinsic motivation and reducing reliance on external aids【9】【10】.

Educators play a crucial role in fostering integrity by modeling the ethical use of AI in their teaching. Demonstrating how AI can be responsibly leveraged—such as for brainstorming or automating routine tasks—helps students understand its appropriate applications. Additionally, discussions about the ethical implications of AI encourage reflection on the consequences of misuse, reinforcing the importance of integrity 【11】 【12】.

The evolving role of AI in education calls for strategies that emphasize understanding and responsibility rather than restriction. By embedding academic integrity within active learning and ethical practice, institutions can move beyond integrity statements to build a resilient academic culture. This approach ensures that students not only comply with ethical standards but also develop critical thinking and decision-making skills essential for an AI-driven future 【6】 【12】.

1.4 Guiding Principles: Learning Through, Not Against AI

Integrating AI into education requires shifting from defensive strategies to proactive frameworks that encourage learning through AI. Educators often perceive AI as a threat to academic integrity, prompting restrictive measures that stifle innovation. However, adopting guiding principles that leverage AI's strengths while addressing its limitations can foster deeper engagement and critical thinking in students 【3】 【6】.

One essential principle is designing assignments that engage students at higher levels of cognitive complexity. Tasks focused on analysis, evaluation, and creation encourage deeper learning by requiring students to interpret context, apply judgment, and generate original responses. For example, educators might ask students to critique the quality of AI-generated text using classroom rubrics or compare its content to human-written work to determine its accuracy and relevance. These assignments support the Analyze and Evaluate levels of Bloom's Taxonomy by prompting students to detect patterns, interpret meaning, and assess decisions using reflective judgment 【8】 【9】. Based on the updated framework, students at the Create level can use AI to support initial brainstorming while still being responsible for developing solutions grounded in ethical reasoning, emotional awareness, and social context—areas where human cognition remains essential. 【13】

Transparency about AI's role in the learning process is equally important. Students should clearly understand expectations for using AI tools and the ethical considerations involved. Discussions on responsible AI use, along with explicit guidance on integrating AI into assignments, equip students to make informed decisions. Studies show that such transparency fosters trust and encourages students to view AI as a resource rather than a shortcut【9】【12】

AI's collaborative potential also creates unique learning opportunities. By framing AI as a partner in problem-solving, educators can design activities that mirror real-world applications. For instance, using AI to analyze large datasets or generate initial ideas for projects prepares students for tasks they may encounter professionally. Research underscores the value of these experiences in developing technological literacy for an AI-driven world【3】【10】.

Finally, adaptability in teaching practices is critical. As AI technologies evolve, educators must remain agile, continually updating methods to reflect advancements and their implications for learning. Staying informed about emerging technologies and engaging in professional development ensures that teaching strategies remain relevant and effective【6】【11】.

Focusing on these principles helps educators develop a balanced framework for integrating AI into education. This approach emphasizes critical engagement, ethical practices, and skill development, enabling students to use AI for innovation and problem-solving. Such alignment enriches the learning experience and prepares students to navigate the complexities of an AI-driven future.【9】【12】.

1.5 Why This Approach Matters

Adopting a framework that positions AI as an educational ally is not just a response to technological advancements—it is essential for preparing students for the demands of an increasingly AI-driven world. This approach prioritizes the development of critical thinking, ethical reasoning, and technological literacy, all crucial for success in both academic and professional environments. Research underscores the urgency of equipping students with these skills as industries increasingly depend on AI for innovation, problem-solving, and decision-making 【3】【6】.

One of AI's greatest educational benefits is its capacity to facilitate personalized learning in ways traditional methods often cannot. For example, AI can analyze individual learning patterns, pinpoint areas for improvement, and provide tailored resources. These tools allow educators to focus on fostering creativity and deeper engagement while addressing students' unique challenges. Studies indicate that personalized learning experiences enabled by AI significantly enhance student outcomes and satisfaction 【8】【9】.

Embracing AI also fosters a culture of innovation and adaptability. By encouraging students to use AI as a tool for exploration and problem-solving, educators equip them to handle complex, real-world challenges. Assignments that require students to use AI for tasks such as data analysis or project planning simulate professional scenarios. This experiential learning not only strengthens technical skills but also builds students' confidence in their ability to navigate an AI-integrated world 【6】【10】.

Equally important is the focus on ethical engagement with AI. Students must learn to recognize both AI's potential and its limitations, as well as its the concerns raises around privacy, bias, and decision-making. Discussions on responsible AI use encourage students to reflect on these societal and ethical issues, fostering accountability and informed decision-making. Research highlights the value of these conversations in developing students' capacity to use AI thoughtfully and responsibly 【9】【12】.

Moreover, integrating AI as an ally in education addresses broader systemic challenges, including the need for scalable and efficient teaching methods. By automating administrative tasks, providing real-time feedback, and supporting differentiated instruction, AI frees educators to focus on high-impact teaching strategies. This comprehensive approach enhances the learning experience and ensures that education systems can meet the evolving needs of students and society 【7】【11】.

Ultimately, this approach prepares students not only to leverage AI's potential but also to use it responsibly for the betterment of society 【6】【12】. By fostering innovation, ethical reasoning, and adaptability, educators can cultivate a generation of learners who thrive in a world where AI is pervasive.

1.6 Chapter 1 Summary

Artificial intelligence has the potential to transform learning at every educational level by supporting personalized, adaptable, and innovative approaches. In primary and secondary education, AI can help create individualized learning paths to enhance foundational skills and address specific student needs. In higher education, AI streamlines research processes, provides real-time feedback, and supports experiential learning. These applications prepare students for professional environments where technology plays a critical role in decision-making and problem-solving. Thoughtful AI integration bridges the gap between theoretical knowledge and practical application, fostering both academic growth and career readiness【3】【6】.

This chapter has emphasized the importance of viewing AI as an ally in education rather than a source of apprehension. Shifting perspectives requires designing tasks and learning environments that encourage ethical engagement, critical thinking, and technological literacy. Educators must lead by modeling responsible AI use, offering clear guidance, and encouraging students to explore AI's capabilities while recognizing its limitations. These efforts position AI as a collaborative tool that enriches education rather than undermining it. This approach equips students with the skills and ethical frameworks necessary to thrive in an increasingly AI-integrated world【8】【9】.

As AI evolves, its role in education must continually be reassessed and adapted to align with societal and technological shifts. Institutions should cultivate a culture of continuous innovation by encouraging educators and students to explore AI's potential. This approach requires ongoing professional development for educators, collaborative exploration opportunities, and integrating AI tools into everyday learning scenarios. These practices ensure that AI becomes a resource for enhancing equity, accessibility, and learning outcomes for diverse student populations. This adaptability ensures that the integration of AI remains relevant and effective over time【10】【12】.

Preparing students for an AI-driven future involves more than providing access to tools—it requires teaching them how to use those tools responsibly and effectively. Assignments that emphasize real-world applications, critical analysis, and ethical reasoning equip students for both the opportunities and challenges AI presents. These experiences not only develop technical proficiency but also instill a sense of accountability, empowering students to use AI for positive contributions to society. By embedding these principles into education, AI becomes a powerful catalyst for growth and innovation, shaping learners ready to excel in any field they choose 【6】【9】.

Integrating AI across educational levels ensures that students are prepared for the realities of a rapidly changing world. This chapter underscores the importance of intentionality, adaptability, and ethical responsibility in using AI as a resource. These principles enhance current learning experiences and lay the groundwork for students to become leaders in an AI-driven society. Through thoughtful design and implementation, AI can enrich education at all levels, creating opportunities for growth, innovation, and lifelong learning 【9】【12】.

References

1. Anderson, J., & Miller, K. (2021). *AI-enhanced learning: Designing assignments that engage and challenge students.* Computers in Education, 85, 102-118.
2. Greenfield, C. (2022). *Plagiarism in the age of AI: Addressing academic integrity through engagement.* Academic Integrity Journal, 9(4), 223-240.
3. Nguyen, H., & Brown, T. (2018). *Teaching for the future: Leveraging AI for critical thinking and problem-solving.* International Journal of Educational Development, 63, 112-128.
4. Thomas, L. (2021). *The role of AI in transforming education: A global perspective.* World Education Review, 29(2), 134-150.
5. Allen, P., & Rivera, G. (2023). *Preparing students for an AI-driven world: Educational best practices.* Future of Learning, 12(1), 89-103.
6. McArthur, D., & Lewis, M. (2020). *Revolutionizing education with AI: Opportunities and challenges.* Journal of Educational Technology, 17(3), 45-57.
7. Smith, T., & Johnson, R. (2019). *Strategies for integrating AI in the classroom: Overcoming resistance to innovation.* Educational Research Quarterly, 42(2), 14-29.
8. Anderson, J., & Miller, K. (2021). *AI-enhanced learning: Designing assignments that engage and challenge students.* Computers in Education, 85, 102-118.

9. Greenfield, C. (2022). *Plagiarism in the age of AI: Addressing academic integrity through engagement.* Academic Integrity Journal, 9(4), 223-240.
10. Nguyen, H., & Brown, T. (2018). *Teaching for the future: Leveraging AI for critical thinking and problem-solving.* International Journal of Educational Development, 63, 112-128.
11. Thomas, L. (2021). *The role of AI in transforming education: A global perspective.* World Education Review, 29(2), 134-150.
12. Allen, P., & Rivera, G. (2023). *Preparing students for an AI-driven world: Educational best practices.* Future of Learning, 12(1), 89-103.
13. Zaphir, L., Lodge, J. M., Lisec, J., McGrath, D., & Khosravi, H. (2024). How critically can an AI think? A framework for evaluating the quality of thinking of generative artificial intelligence (Version 2.0) [White paper]. arXiv. **https://arxiv.org/abs/2406.14769**

Chapter 2 - Understanding Probability and Specificity in AI

Probability and specificity significantly influence effective artificial intelligence (AI) utilization in education. Medical assessments frequently use probability and specificity to inform clinical decision-making processes. Working as a paramedic illustrates the importance of probability clearly. For example, when responding to a patient with abdominal pain, a paramedic immediately eliminates diagnoses such as pregnancy or ovarian cysts if the patient is male, assigning these conditions a probability of zero. Once improbable conditions are excluded, specificity guides further diagnostic decisions. Determining the precise location, type, or radiation of pain helps paramedics narrow down possible conditions. Pain on the right side suggests possibilities like appendicitis or right kidney issues, while pain radiating to the left shoulder could indicate gallbladder disease, and a tearing sensation between the shoulder blades could suggest a dissecting aneurysm. Although specificity may not provide absolute certainty, it accelerates the diagnostic process.

AI systems operate through comparable mechanisms of probability and specificity. AI systems assess the probability of correct responses based on input specificity. Generic prompts often result in broad and imprecise AI outputs. Prompts demanding detailed, context-specific information compel AI systems to generate precise and accurate responses. Educators who understand these principles design assignments that reduce ambiguity and enhance clarity.

Effectively utilizing AI in educational settings requires educators to understand probability and specificity implications fully. Detailed and precise assignment prompts enhance AI response accuracy, supporting student engagement and authentic critical thinking. Assignments requiring personalized reflections, specific analytical content, and precise contextual details reduce generic AI-generated responses. Educators equipped with an understanding of probability and specificity improve assignment design, optimizing AI capabilities to promote genuine student participation and deeper learning.

The role of artificial intelligence (AI) in education extends beyond facilitating learning; it also involves understanding how AI operates and applies knowledge. Two fundamental principles driving AI functionality, particularly in decision-making and predictive analytics, are probability and specificity. Probability helps AI systems assess the likelihood of events or outcomes based on input data, while specificity ensures tailored responses to context-specific queries. These principles are crucial for educators and students to grasp as they integrate AI into the learning process 【1】【2】.

Understanding probability in AI requires analyzing how systems detect data patterns to predict outcomes with varying degrees of certainty. For instance, adaptive learning platforms use probability to identify the most suitable learning path for students based on their performance history. Specificity refines this process by tailoring predictions to the user's unique needs, enhancing the accuracy and relevance of AI-generated responses. Research shows that combining these principles can significantly improve the effectiveness of AI tools in educational settings 【3】【4】.

Educators play a vital role in teaching students how to work with AI systems based on probability and specificity. By introducing tasks that require evaluating AI outputs or analyzing its decision-making process, students develop technological literacy and strengthen critical thinking and problem-solving skills. Integrating these concepts into curricula equips students with the knowledge necessary to navigate and utilize AI effectively in both academic and professional contexts 【5】【6】.

2.1 The Mechanics of Probability in AI

Probability is central to how AI evaluates and responds to prompts. AI systems rely on probability to predict the most likely answer based on available data. For example, if a student asks, "Is the sky orange today?" the AI might consider factors such as the time of day, weather conditions, and geographic location. The likelihood of the sky being orange at midday is low, but it increases during sunrise or sunset when orange hues are common. Probability allows AI to generate responses based on patterns, although ambiguous prompts can lead to errors. Crafting precise questions reduces this margin of error and highlights the importance of specificity in improving AI outputs 【1】【2】.

In healthcare, probability-driven AI models calculate disease likelihood based on patient data, comparing individual cases to extensive medical datasets. These models can predict heart disease risks or detect early signs of diabetes by evaluating variables such as blood pressure, cholesterol levels, and lifestyle factors. This application illustrates probability's broader role in predictive insights. Educators can leverage this understanding to design assignments that challenge students to engage critically, rather than rely solely on AI for solutions 【3】【5】.

Understanding probability in AI also informs the creation of assignments and rubrics that promote active student engagement. Educators might develop tasks requiring students to evaluate AI-generated outputs or analyze the probabilities behind specific predictions. For example, an assignment could ask students to critique an AI-generated response, examining how probabilities were calculated and identifying factors affecting prediction accuracy. Integrating these concepts into rubrics ensures that students are assessed on their critical thinking and problem-solving skills rather than their ability to use AI passively. This approach fosters a deeper understanding of AI's capabilities and limitations 【5】【6】.

2.2 The Role of Specificity in Enhancing Learning and Diagnostics

Specificity directly affects how AI interprets prompts and how students engage with learning tasks. As discussed in Section 2.1, a vague question such as "Is the sky orange today?" typically produces an unclear or noncommittal answer, since the AI has no contextual information to analyze. Adding key variables—like time of day, geographic location, and weather conditions—significantly improves the relevance and precision of the response. A revised prompt such as "Is the sky orange at 7:00 AM on a bright sunny day in Carlisle, Ohio?" allows the AI to access known sunrise patterns and environmental cues, resulting in a more accurate answer. This type of refinement encourages students to revise and reframe their questions actively, fostering critical thinking and a deeper understanding of how input quality affects AI output 【7】【8】.

The same principle applies to rubric design. When rubrics require students to include specific details—such as location, time, or weather conditions—they may inadvertently guide students to construct highly specific prompts that AI can answer with minimal human input. For example, a rubric that directs students to describe the sky at "7:00 AM on a bright sunny day in Carlisle, Ohio" enables AI to produce a complete and contextually appropriate response with little effort from the student. By contrast, rubrics that leave room for interpretation encourage students to make independent decisions about context and meaning. These more open-ended designs prompt deeper engagement, as students must identify which details are relevant and how to incorporate them effectively into their work.

Consider two versions of a rubric for a short writing task. The first rubric reads: "Describe a morning scene in Carlisle, Ohio, at 7:00 AM on a sunny day. Your response must include sensory details about the weather, surroundings, and time of day." This version allows students to input (copy and paste) those exact criteria into AI, often resulting in generic yet acceptable responses that lack personal insight. The second rubric shifts responsibility to the student: "Describe a meaningful morning scene in a location that matters to you. Include details that reflect the time of day, weather, and personal significance." In this case, the student must select a setting, define personal emotion, and draw from personal experience—requirements that AI cannot fulfill without human input. Moving from rigid to flexible criteria increases student engagement and reduces the effectiveness of AI as a shortcut, ultimately promoting more meaningful engagement with the assignment.

Instructors must carefully determine when to include or exclude specificity in assignments and rubrics to balance learning objectives with effective AI use. Including specificity can improve clarity and accuracy but may reduce critical engagement if students rely solely on AI for fully specified prompts. In contrast, omitting specificity encourages students to identify missing elements and refine their queries, fostering active participation and critical thinking. For example, asking students whether the sky could appear orange on a bright sunny day in Carlisle, Ohio, at a given time—without specifying additional details—requires them to apply contextual understanding and reasoning. Striking this balance allows educators to design assignments that promote meaningful learning while leveraging AI as a tool for exploration and insight 【9】【10】.

2.3 Practical Applications of Probability and Specificity in Education

Assignments that incorporate probability and specificity help move students beyond surface-level thinking by requiring them to make informed, analytical decisions. For example, a high school assignment might ask students to evaluate the causes of the American Revolution in relation to economic policies. Without specifying a region or policy type, students must determine the scope of their analysis. This prompts deeper engagement, as they might focus on New England's tax resistance, British trade restrictions, or colonial responses to mercantilist laws. The assignment design demands that students identify relevant variables, weigh their significance, and develop arguments based on historical context and available evidence.

In secondary education, real-world scenarios can further integrate probability and specificity to promote critical reasoning. A history project could ask students to predict the potential outcomes of diplomatic negotiations during World War II based on the known alliances and prevailing economic pressures of the time. A strong rubric might require the inclusion of relevant countries, political motivations, and strategic interests. These components compel students to apply contextual knowledge and use evidence to support probabilistic conclusions. The complexity of the task discourages passive summarization and encourages original interpretation supported by both primary and secondary sources.

Effective assignment design also relies on rubrics that assess a student's ability to contextualize information and refine their reasoning. A geography assignment might ask why specific regions emerged as major trade hubs during the 18th century. Rather than listing locations, students must analyze factors such as geographic accessibility, proximity to maritime routes, and the influence of local resources or colonial infrastructure. A rubric that emphasizes variable comparison, pattern recognition, and significance of detail helps ensure that student responses reflect thoughtful engagement. At the same time, educators must be careful not to overload rubrics with so much specificity that students default to replicating patterns rather than developing their own conclusions.

When probability and specificity are integrated into both assignment prompts and assessment rubrics, students are required to make intentional academic decisions. Instead of relying on generalizations, they must consider which details matter and why, forming a habit of inquiry that encourages original thought. These tasks teach students to evaluate uncertainty, narrow scope appropriately, and support conclusions with targeted evidence. Assignments designed with these principles discourage reliance on AI-generated summaries by introducing complexity that requires student judgment. Over time, these strategies help students build the analytical habits necessary for real-world problem solving and academic independence.

2.4 Building Assignments with Probability and Specificity

Probability and specificity underpin assignments that prepare students to identify and apply patterns effectively. Probability-based tasks encourage students to evaluate multiple possible outcomes, fostering critical thinking. For instance, a mathematics assignment might ask students to predict the results of a probability experiment, such as rolling dice or analyzing statistical trends in data sets. Specificity ensures students work with precise variables, such as calculating the likelihood of a specific sum appearing based on given probabilities. This balance between open-ended exploration and defined constraints promotes analytical reasoning.

Specificity also helps students transition from recognizing patterns to applying them in complex scenarios. An assignment focused on patterns in literature might require students to analyze a recurring motif across chapters and connect it to broader themes. By specifying details such as character interactions or particular narrative moments, students are guided to actively discover connections rather than rely on generalized observations. Rubrics can evaluate how well students use probabilistic reasoning to predict plot developments and analyze specific textual elements to support their conclusions.

Upcoming chapters will delve into how pattern recognition strengthens assignment design, providing frameworks for creating tasks that challenge students to synthesize and apply their understanding. The concepts of probability and specificity outlined in this chapter serve as tools for crafting assignments that demand deeper engagement. As readers progress, they will explore how these foundational principles are integral to constructing assignments that teach content and develop higher-order cognitive skills. Educators can foster exploration and precision in student learning by incorporating probability and specificity into assignment structures, setting the stage for advanced educational practices.

2.5 Transitioning from Probability and Specificity to Assignment Design

Understanding how probability and specificity influence assignment outcomes lays a foundation for designing effective tasks. The principles discussed in this chapter guide educators in structuring assignments that prompt critical thinking, require detailed responses, and encourage meaningful engagement. Future chapters will build on these ideas by integrating pattern recognition into broader assignment frameworks. Probability helps students assess likely outcomes based on data, while specificity keeps assignments grounded in context, requiring active student input rather than passive reliance on AI.

Educators can use probability and specificity to transition from theoretical concepts to practical applications. For instance, a history assignment might analyze probabilistic trends in migration patterns while incorporating specific time frames and regional data. This approach ensures students cannot rely on generalized answers but must interpret specific variables to provide thoughtful responses. Rubrics for such tasks can highlight the importance of detail and probabilistic reasoning, evaluating students' ability to synthesize and apply knowledge critically.

Upcoming chapters will provide actionable strategies for designing assignments that emphasize student engagement and critical thinking. The next focus on pattern recognition will explore how identifying and connecting recurring elements across disciplines enhances learning. The interplay between probability, specificity, and pattern recognition will be examined to offer a cohesive framework for crafting assignments that foster meaningful learning experiences and prepare students for academic and real-world challenges.

2.6 Bridging Probability, Specificity, and Pattern Recognition

Probability and specificity form the instructional foundation behind the kind of assignments I have spent my career designing—assignments that demand more than memorization. These principles mirror the same reasoning I used when building simulation scenarios that required students to think through uncertainty and act with intention. Probability encourages students to evaluate possible outcomes, weigh evidence, and consider alternatives. Specificity, in turn, forces clarity—pushing students to narrow their focus, ground their responses in context, and avoid vague generalities. Together, these elements move students from passive observers to active participants, creating opportunities where thinking is not optional. I wrote this book because too many educational practices still reward surface-level completion. My goal is to help educators design tasks that require students to interpret, respond, and reflect—skills AI cannot replicate without being told exactly what to say.

The next chapter builds on this foundation by examining how pattern recognition can enhance critical thinking and strengthen assignment design. Pattern recognition helps students identify recurring ideas, connect knowledge across subjects, and synthesize meaning from complex or incomplete information. When combined with probability and specificity, pattern recognition becomes a powerful instructional tool—one that fosters deeper cognitive processing and authentic learning. This next section continues the broader aim of this book: to offer clear, specific strategies for building assignments that challenge students to think deeply, engage honestly, and develop the intellectual independence they need in an AI-shaped world.

2.7 Chapter 2 Summary

Chapter 2 introduces the foundational concepts of probability and specificity as they apply to the design of assignments in an AI-influenced educational landscape. Drawing on parallels from clinical reasoning in medical settings, the chapter explains how AI systems rely on probabilistic modeling to generate responses and how specificity in prompts increases the accuracy and depth of those responses. Educators are encouraged to frame prompts that demand precise, context-rich input—thus requiring students to engage with course material meaningfully rather than defaulting to generic AI outputs. Assignments designed around these principles not only improve the relevance of AI-generated responses but also promote deeper student reasoning, critical reflection, and personalized learning outcomes.

The chapter also explores how these concepts translate into practical classroom strategies, such as crafting rubrics and tasks that balance clarity with open-ended reasoning. Examples from secondary education highlight how students must identify relevant variables, anticipate outcomes, and justify conclusions—skills that align with real-world problem solving. By embedding probability and specificity into the structure of assignments and assessments, educators shift the burden of critical decision-making back to the student. The chapter concludes by linking these principles to the development of pattern recognition skills, setting the stage for the following chapter on designing assignments that promote higher-order thinking and intellectual independence.

References

1. Russell, S., & Norvig, P. (2021). *Artificial intelligence: A modern approach* (4th ed.). Pearson Education.
2. Murphy, K. P. (2012). *Machine learning: A probabilistic perspective.* MIT Press.
3. Goodfellow, I., Bengio, Y., & Courville, A. (2016). *Deep learning.* MIT Press.
4. Chollet, F. (2018). *Deep learning with Python.* Manning Publications.
5. Bishop, C. M. (2006). *Pattern recognition and machine learning.* Springer.
6. LeCun, Y., Bengio, Y., & Hinton, G. (2015). Deep learning. *Nature*, 521(7553), 436-444. https://doi.org/10.1038/nature14539
7. Silver, D., Schrittwieser, J., Simonyan, K., Antonoglou, I., Huang, A., Guez, A., ... & Hassabis, D. (2017). Mastering the game of Go without human knowledge. *Nature*, 550(7676), 354-359. https://doi.org/10.1038/nature24270
8. Topol, E. J. (2019). *Deep medicine: How artificial intelligence can make healthcare human again.* Basic Books.
9. Esteva, A., Kuprel, B., Novoa, R. A., Ko, J., Swetter, S. M., Blau, H. M., & Thrun, S. (2017). Dermatologist-level classification of skin cancer with deep neural networks. *Nature*, 542(7639), 115-118. https://doi.org/10.1038/nature21056
10. Johnson, K. W., Torres Soto, J., Glicksberg, B. S., Shameer, K., Miotto, R., Ali, M., & Dudley, J. T. (2018). Artificial intelligence in cardiology. *Journal of the American College of Cardiology*, 71(23), 2668-2679. https://doi.org/10.1016/j.jacc.2018.03.521

Chapter 3: Building Assignment for Meaningful Learning

Recognizing patterns is a crucial educational skill that enables students to connect concepts, identify trends, and synthesize information across disciplines. Integrating pattern recognition into assignment design fosters critical thinking and analytical reasoning. Assignments that focus on patterns encourage students to move beyond rote memorization, prompting them to find meaningful connections in the material they study. Whether identifying recurring themes in literature, analyzing trends in historical events, or mapping relationships in scientific data, pattern recognition empowers students to engage deeply with their learning and apply insights to broader contexts.

This chapter explores strategies for designing assignments that incorporate pattern recognition to develop higher-order cognitive skills. Through thoughtful assignment structures, students are encouraged to identify and apply patterns while leveraging the probability and specificity principles introduced earlier. By building tasks that challenge students to engage actively with content, educators can cultivate critical thinking and problem-solving abilities. The strategies outlined here will offer actionable steps for creating assignments that inspire meaningful engagement and prepare students for academic and real-world challenges.

Throughout this chapter, probability, specificity, and AI response quality will be used as key indicators to evaluate assignment design. Probability refers to the likelihood that AI can generate a correct or complete answer based on the information provided. Specificity reflects the level of detail or context required in a student's response or prompt. AI dependency is shaped by the relationship between the two. When specificity is low and the probability of AI generating a correct answer is high, students can rely heavily on AI, often bypassing meaningful engagement. Conversely, when an assignment demands high specificity—especially when tied to personal experience, classroom discussion, or localized knowledge—the probability of AI generating a useful response drop, and student engagement increases. Each example in this chapter will use these variables to illustrate how thoughtful assignment design can reduce overreliance on AI and foster critical thinking.

3.1 Designing Assignments to Minimize AI Dependency

Effective assignment design plays a vital role in fostering students' ability to think critically, creatively, and deeply while minimizing dependence on AI-generated outputs. Advanced AI systems such as ChatGPT, Claude AI, and Bard excel at producing comprehensive but often shallow responses to generalized prompts. These outputs may appear sufficient but lack the depth and originality needed for meaningful learning. When assignments do not challenge students beyond surface-level comprehension, engagement tends to diminish.

Reimagining assignments by incorporating specificity, emphasizing the learning process, and applying contextual elements ensures that students engage actively and critically with tasks. Shifting from generalized to tailored prompts encourages deeper exploration and critical thinking. For example, instead of asking, "Describe the benefits of renewable energy," educators could frame prompts that require analytical responses, such as "Evaluate the feasibility of implementing solar energy solutions in urban areas with limited rooftop space."

This chapter offers practical strategies to create assignments that reduce overreliance on AI, balance its use as a supportive tool, and foster authentic intellectual growth.

3.2 The Foundation of AI-Resistant Assignments

AI systems like ChatGPT, Claude AI, and Gemini rely on pattern recognition and probability to generate responses, making them particularly adept at addressing broad or predictable prompts. For instance, a prompt such as *"Discuss the benefits of renewable energy"* allows AI to produce well-structured but generalized answers with minimal effort or student input. While these outputs may provide a basic understanding, they do little to foster deeper engagement or critical thinking.

By contrast, assignments that incorporate specificity and contextual relevance challenge AI's capabilities and require students to provide personal insights, localized information, or detailed analysis. For example, a prompt like *"Analyze the challenges of implementing wind energy in rural areas with limited infrastructure"* demands the integration of classroom knowledge, external research, and contextual reasoning, making it difficult for AI to complete the task independently. The more targeted assignments encourage students to engage with material thoughtfully, develop their perspectives, and cultivate problem-solving skills, thereby strengthening the learning experience.

Table 3.2.1 Examples of AI-Resistant Prompts and AI Responses

Assignment Prompt	Specificity Level	AI Dependency	Example AI Chat GPT Response
"Describe your personal experience with online learning."	High	Low	"I don't have personal experiences, but online learning can vary based on individual preferences."
"Compare local and global policies on renewable energy."	High	Low	"Local policies may focus on community needs, while global policies aim for broader sustainability goals."
"Design a community engagement plan for recycling."	High	Low	"A community engagement plan might include education campaigns and incentives for increased participation."

Table 3.2.1 examples highlight specificity's essential role in reducing AI's ability to independently generate complete and meaningful responses. By requiring personal experiences or localized contexts, such prompts introduce elements that AI cannot easily replicate, compelling students to engage more deeply with the material. This approach fosters active student participation, critical thinking, and the ability to connect concepts to real-world applications. Assignments designed with these principles promote deeper understanding and encourage students to develop original insights, enhancing the educational experience.

3.3 Strategies for Crafting AI-Resistant Prompts

Creating prompts that minimize reliance on AI tools while fostering student engagement requires thoughtful strategies. Educators can encourage deep learning by incorporating personalization, contextual relevance, process-oriented tasks, and real-world applications. These strategies help maintain assignment rigor while promoting critical thinking, creativity, and collaboration at all educational levels—from elementary school to higher education. Below are six key strategies, each supported by examples and tailored to different academic stages.

Educators seeking a quick visual reference can refer to ***Appendix A: AI-Generated vs. Student-Tailored Prompts.*** This tool compares common assignment formats and highlights how intentional design choices can reduce AI dependency and deepen student engagement.

3.4 Personalization and Contextualization

Assignments that require students to draw from personal experiences or local contexts support deeper learning and reduce the usefulness of AI-generated content. When students are prompted to include specific, individualized details, they must reflect, select meaningful examples, and construct responses grounded in experience—tasks that generative AI cannot complete independently. Personalization introduces cognitive and contextual variables that do not exist in public datasets or predictive models. These assignments encourage original thinking and align with the higher-order domains of Bloom's Taxonomy, particularly the *apply*, *analyze*, and *evaluate* levels【1】.

This level of cognitive engagement is especially important in adult education. Learners bring prior experiences and social context to each learning task, and when instructional design honors that reality, students are more likely to connect with content meaningfully. Carol Dweck's work on growth mindset reinforces the value of these types of assignments. When students are invited to incorporate their own lives into academic tasks, they are more likely to view learning as a process shaped by effort and self-awareness rather than fixed ability【2】. Encouraging students to explore their own perspectives builds a classroom culture that supports motivation, reflection, and intellectual risk-taking.

Mary-Ann Winkelmes' Transparency in Learning and Teaching (TILT) framework also supports this approach. Her research emphasizes the importance of clarifying both the *purpose* and *relevance* of assignments. When students understand why personal context matters and how it contributes to learning goals, they are more likely to engage intentionally and persist through challenging tasks【3】. This is especially significant for first-generation and underrepresented students, who may otherwise misunderstand assignment expectations. Personalization, when paired with transparency, helps to close that gap.

The relationship between specificity and AI dependency becomes especially clear when comparing how different types of assignment prompts function in practice. Prompts that encourage vague or generalized responses are easily handled by AI. In contrast, prompts that require local detail, personal narrative, or lived experience reduce AI effectiveness by demanding information that AI cannot generate without explicit guidance. The examples below show how changes in specificity alter both the level of required thinking and the usability of AI-generated output.

Table 3.4.1 uses real responses from generative AI platforms (ChatGPT and Claude). Each assignment prompt was entered exactly as written, and the response recorded reflects what a student would see when copying and pasting the prompt into the system. While these tools continue to evolve, the core pattern remains consistent: the more an assignment requires individual judgment and context, the less usable the AI response becomes.

Table 3.4.1: Specificity and AI Dependency in Personalization Assignments

Assignment Prompt	Specificity Level	AI Dependency	Example AI Response
Describe your favorite activity from summer break	High	Low	"I can't describe personal activities, but many students enjoy swimming and outdoor games." (ChatGPT)
Analyze how a community project impacted your neighborhood	High	Low	"Community projects such as planting trees often bring residents together and beautify the area." (Claude)
Reflect on how an internship shaped your professional development	High	Low	"Internships are valuable for gaining hands-on experience and applying classroom knowledge." (ChatGPT)
Discuss the benefits of teamwork	Low	High	"Teamwork helps people collaborate effectively, solve problems, and reach shared goals." (Claude)
Explain the importance of communication in the workplace	Low	High	"Communication in the workplace ensures tasks are completed efficiently and misunderstandings are avoided." (ChatGPT)

These examples confirm that when students are required to add localized, personal, or reflective context, AI tools produce responses that lack substance or relevance. In contrast, general prompts lead to full responses that appear complete but involve no authentic cognitive effort from the student. Personalization not only challenges students to think more deeply but also shifts the purpose of the task from completion to reflection.

By combining personal context, structured reflection, and transparent expectations, educators can design assignments that support intellectual development while reducing AI dependency. These practices promote the types of cognitive, emotional, and ethical growth that AI cannot simulate. When educators incorporate principles from Bloom, Dweck, and Winkelmes, they create learning environments where students are expected to think—and given every reason to rise to that challenge.

3.5 Process-Oriented Assignments

Process-oriented assignments require students to document their reasoning and decision-making steps, shifting the focus from simply presenting outcomes to explaining how conclusions are reached. Structure like this makes AI-generated solutions less effective, as success depends on the student's ability to articulate process, logic, and reflection.

For instance, an elementary student might be asked to solve a math problem and explain each calculation step. A high school student may outline their scientific reasoning in a lab report, while a medical student could document the clinical thinking that led to a diagnosis. Each of these tasks demands explanation, justification, and structure—elements that AI cannot supply without specific guidance from the user.

Table 3.5.1 shows how varying levels of specificity in prompts influence AI dependency. Tasks that are overly general allow AI to produce full answers without student engagement. In contrast, detailed, process-focused prompts require input that AI cannot generate independently, ensuring students remain at the center of the learning process.

Table 3.5.1: Process-Oriented Assignments

Assignment Prompt	Specificity Level	AI Dependency	Example AI Response
Solve 12 × 7 and explain your steps.	Low	High	The product is 84, calculated by adding 12 seven times.
Explain the clinical reasoning process for chest pain from the case presented in class.	High	Low	A chest pain diagnosis might involve checking vitals and considering heart or lung conditions, based on what was observed during the classroom case discussion.

The first assignment, 'Solve 12 × 7 and explain your steps,' has a low level of specificity. This allows AI to answer with near-perfect probability, as the question is simple, direct, and based on well-established facts. The probability of AI producing an incorrect response is close to zero, making it highly accessible and reliable for AI tools. As a result, students can depend on AI to complete the task with minimal personal input, leading to a high level of AI dependency.

The second assignment, 'Explain the clinical reasoning process for chest pain from the case presented in class,' requires the student to incorporate details from a specific classroom scenario. AI tools lack access to classroom discussions and cannot predict those unique elements with accuracy. The probability of AI generating a correct and complete response is significantly lower unless the student provides additional specificity within the prompt. This need for contextual information shifts the cognitive responsibility back to the student and reduces their ability to rely solely on AI.

3.6 Real-World Applications

Assignments based on real-world scenarios or project-based learning require students to analyze variables, synthesize information, and apply knowledge creatively. This strategy fosters practical application and reduces reliance on AI, as students must provide unique insights and conduct environment-specific research.

For instance, high school students might design a marketing plan to promote environmental awareness at their school. Business school students could draft financial recovery strategies for a fictional small business, incorporating tailored solutions based on market dynamics.

Table 3.5.1: Specificity and AI Dependency in Personalization Assignments

Assignment Prompt	Specificity Level	AI Dependency	Example AI Response
"Design a marketing plan for high school art events."	High	Low	"You could use social media to promote events and offer incentives like free tickets to students."
"Develop a financial recovery plan for a small business."	High	Low	"The plan could include reducing expenses and diversifying income streams."

3.6 Collaborative and Interactive Tasks

Collaborative tasks rely on human interaction and group dynamics, making them difficult for AI to replicate effectively. These assignments promote teamwork, communication, and negotiation while ensuring originality through diverse perspectives.

Elementary students could work in pairs to create and illustrate a short story. For instance, they might be given the prompt, "Work in pairs to create a story about teamwork and illustrate it." The specificity level of this task is high, and AI dependency is low. While AI might respond with a generic example such as, "A story about teamwork could involve friends helping each other to win a race," it cannot engage in the collaborative dialogue or creative illustration required by the task.

High school students might collaborate on designing a sustainable school garden. An assignment like "Collaborate on a project to design a sustainable school garden" also has high specificity and low AI dependency. An AI-generated response may suggest, “A sustainable garden might include planting drought-resistant plants and composting.” Although technically correct, such a response lacks the depth and shared planning that occurs through genuine student collaboration.

Business school students could prepare a joint presentation analyzing renewable energy trends, combining financial, environmental, and policy perspectives. Like the earlier examples, the success of this task depends on multiple viewpoints and the ability to integrate discussion into a coherent product. These collaborative assignments reduce AI dependency by prioritizing shared thinking and real-time interaction.

3.7 Referencing Classroom Materials

Assignments requiring students to draw from classroom discussions, lectures, or specific course materials ensure responses remain unique to the learning environment.For instance, elementary students might recount a story discussed during a class reading session. High schoolers could critique a historical event using lecture notes, while medical students evaluate case studies presented during their clinical rotations, incorporating details accessible only to classmates.

Table 3.7.1: Material Referencing Example:

Assignment Prompt	Specificity Level	AI Dependency	Example AI Response
“Summarize the story we read in class and explain its main message.”	High	Low	"Stories often have messages like the importance of kindness or perseverance."
“Analyze the key points from today’s lecture on cardiovascular health.”	High	Low	"Cardiovascular health involves maintaining a healthy diet and regular exercise."

3.8 Multilayered Assignments

Multilayered assignments are designed to help students connect multiple concepts, build on earlier activities, and synthesize what they’ve learned over time. These tasks promote iterative learning, where students return to and refine their thinking as they progress through an assignment. Educators use these assignments to reinforce critical thinking by challenging students to apply knowledge in increasingly complex ways. Students must revisit earlier steps, reflect on new information, and adjust their responses accordingly. These assignments help learners develop habits of revision, analysis, and deeper engagement.

An example at the elementary level involves a plant observation project that spans multiple days. Students identify three plants in their neighborhood, document how each one grows, and record observations over time. They must describe environmental influences such as light, water, or soil conditions, and compare plant development under different circumstances. Each step builds on the previous one, encouraging careful observation and longitudinal thinking. AI cannot replicate these personal, localized insights or mimic changes observed in real time.

In a medical education context, students may work through a multi-stage case study involving a patient with chest pain. Initially, students propose a diagnosis based on symptoms and basic history. As the case unfolds, new diagnostic data is introduced, requiring students to reassess and update their treatment plan. Each decision must be justified, and changes must reflect logical reasoning based on new evidence. The layered nature of the task mimics real-world clinical decision-making and develops diagnostic precision.

AI-generated responses often fall short when applied to assignments that evolve over time. For example, in the elementary plant activity, an AI might respond, "Common plants include grass and trees, which grow with water and sunlight." While accurate, this lacks specificity and fails to reflect real observation or variation over time. Likewise, an AI might suggest "coughing and fever indicate a respiratory infection," but without access to the evolving clinical picture, the response remains generic and incomplete.

Educators should design multilayered assignments with clear stages that require students to reflect, refine, and justify their thinking. Each layer should increase complexity, forcing students to build upon prior knowledge and adapt their ideas. When students document their process and explain revisions, the assignment becomes a record of intellectual growth. These structures reduce AI dependency because the evolving nature of the task cannot be completed in a single step. Multilayered assignments develop essential skills such as adaptability, reasoning, and persistence.

Educators can create AI-resistant prompts that encourage critical thinking and engagement across all educational levels by incorporating strategies such as personalization, process documentation, real-world scenarios, collaboration, referencing class material, and layering tasks. This comprehensive approach ensures that students develop the skills to navigate complex problems while minimizing dependency on AI.

3.9 Chapter 3 Summary

Chapter 3 presents a comprehensive framework for designing assignments that promote meaningful student learning while minimizing reliance on AI-generated responses. The chapter introduces the concepts of probability and specificity as key tools for shaping assignment prompts. When specificity is low and the probability of AI generating a correct answer is high, students can bypass deeper thinking. In contrast, when tasks demand detailed input tied to personal experience, classroom discussion, or local context, the probability of AI responding effectively drops. This relationship becomes the foundation for designing prompts that challenge students to interpret, apply, and synthesize ideas—skills that cannot be outsourced to AI.

To illustrate these principles, the chapter outlines six strategies: personalization, process orientation, real-world application, collaboration, referencing classroom material, and multilayered task design. Each strategy includes examples that show how tailored prompts raise the cognitive demand and reduce the value of generic AI-generated output. Whether students are reflecting on internships, designing community projects, or revising a diagnosis based on evolving case data, the assignment structure demands student thought, context, and input. These instructional designs not only foster engagement but also prepare learners for academic independence and professional problem-solving in environments where AI is present but cannot substitute for human reasoning.

Chapter 4: Detecting AI-Dependent Responses

Detecting reliance on artificial intelligence (AI) in student work is not about catching something entirely new—it is about responding to how cheating has evolved in both scale and subtlety. Academic dishonesty has always existed, whether through peer-written essays, contract cheating, or ghostwritten slides. What sets AI apart is not just accessibility, but invisibility. Tools like ChatGPT and Claude allow students to generate plausible, grammatically clean responses in seconds, often without ever engaging with the assignment prompt or underlying content. The ease of use, paired with rapid output and surface-level coherence, makes these tools especially appealing for students seeking shortcuts. As educators, the challenge is no longer just identifying plagiarism—it is recognizing the absence of learning when the product appears complete.

A clear example of AI's influence has emerged in medical education, particularly in student applications for medical school and residency. Faculty began noticing that personal statements—while grammatically polished—lacked depth, emotion, and narrative authenticity. Many were filled with vague expressions of empathy and generic references to patient care that could apply to almost anyone. Although generative AI tools can fabricate reflective writing that mimics personal insight, these responses often fail to demonstrate the emotional nuance, self-awareness, or coherence found in genuine lived experiences. This became especially evident during interviews. When students were asked to elaborate on the clinical moments they had described, some struggled to recall specific details, demonstrate insight, or express an emotional connection to the scenario. The responses felt disconnected—well-written on paper, but hollow in person.

Faculty were not opposed to students using AI to finesse grammar or improve clarity. In fact, many acknowledged that AI can help refine writing in the same way a peer editor or writing center might. What raised concern was when students used AI to construct the entire narrative, replacing lived experience with generated content. In response, faculty revised the personal statement prompt to require specific clinical experiences and emotionally grounded reflection. These revisions made it more difficult for students to submit content generated entirely by AI without personal involvement. The change not only revealed who had genuinely engaged in clinical learning, but also helped admissions teams differentiate between borrowed narratives and those shaped by experience. Detecting AI use is not about identifying flawed writing—it's about identifying when learning, thinking, and reflection are missing from the student's work.

These patterns in medical education reflect a broader instructional issue: how to distinguish writing that has been reviewed from writing that reflects learning. Chapter 4 continues the discussion of specificity and probability by applying them to the detection of AI-generated responses. Understanding how vague prompts lead to general outputs helps educators identify points where students can bypass the thinking process. When assignments lack clear expectations for reasoning or detail, students can generate complete responses without contributing their own input. Signals such as repetition, vague phrasing, or missing course material help instructors revise assessments to support original responses, context, and cognitive work. Many educators now aim to shift the responsibility for specificity onto the student. Rather than supplying all the details in a prompt, they design tasks that require students to make interpretive decisions, select relevant content, and demonstrate understanding. The goal is not to monitor AI use but to structure assignments in ways that require student contribution, making the thinking process observable and the learning outcome credible.

4.1 Identifying Indicators of AI Dependency in Student Assignments

Educators today face the challenge of navigating a complex landscape where artificial intelligence (AI) plays a significant role in student work. Distinguishing between authentic student contributions and AI-generated content is essential for fostering meaningful engagement. Responses heavily reliant on AI often lack the richness and depth found in assignments grounded in personal reflection, localized knowledge, or classroom-specific insights. For example, a generic topic summary that fails to reference specific course materials or localized examples may signal limited student engagement. Recognizing such patterns helps educators assess whether students are meaningfully interacting with the material or merely using AI tools for surface-level analysis.

Key indicators of AI dependency include generalized phrasing, the absence of critical reasoning, and uniformity in structure that doesn't align with assignment requirements. When assignments call for analysis or interpretation, responses that simply paraphrase widely available information without personal insight may suggest minimal effort. Educators can address these challenges by designing tasks that require student-specific contributions, such as incorporating personal experiences or integrating material from class discussions. These elements limit the effectiveness of AI tools and encourage deeper student engagement 【3】【4】.

Incorporating these insights into assignment design involves adapting tasks and rubrics to prioritize originality and depth. Rubrics should assess the inclusion of unique insights, application of course-specific knowledge, and evidence of critical thinking. Responses that fall short of these criteria can help identify AI-influenced submissions. This approach supports educators in maintaining academic integrity while promoting deeper learning and engagement 【4】【5】.

4.2 Generalized Content

Assignments that do not require students to integrate specific details or contextual elements often produce generalized responses lacking depth and originality. Such content frequently mirrors AI-generated outputs, which draw from vast databases of general information. While these responses may appear coherent and relevant, they rarely engage students in critical thinking or deep analysis. Designing assignments that prompt students to provide unique insights, personal reflections, or localized data can significantly enhance engagement and learning outcomes.

When assignments are tailored to specific prompts requiring personalized input, students actively engage in the learning process. For example, asking students to analyze the impact of a local community event encourages them to incorporate environmental knowledge, historical context, and observed outcomes. AI cannot replicate the nuanced understanding gained through lived experiences or familiarity with classroom discussions. This approach minimizes reliance on AI and promotes student ownership of tasks, increasing intellectual investment and engagement.

Courses that teach students how to use AI responsibly often introduce prompt templates that leave intentional gaps for users to fill in. These gaps, often marked by brackets such as { }, force users to provide specific, original input before the AI can generate a meaningful response. For example, a prompt might read, "Create an action figure using {superpower} and name it {character name}." The AI cannot complete the task until the user adds relevant and personal information. This model works because it shifts the responsibility for specificity to the user. The same principle can be applied to academic assignment design. Instead of offering a fully defined prompt, instructors can design tasks that require students to select a context, insert local data, or reflect on an experience before responding. For instance, rather than asking, "Write about a cultural tradition," the assignment could require, "Describe a cultural tradition celebrated in your community that you have personally observed or participated in." In both cases, the structure blocks AI from generating meaningful content unless the student provides input based on lived experience or unique knowledge. Embedding these requirements into assignments helps ensure the student is not just submitting a response—but participating in its creation.

The following tables illustrate the difference between generalized and context-specific assignments. Table 4.2.1 highlights how the lack of specificity enables AI-generated responses, while Table 4.2.2shows how adapted assignments demand deeper engagement. These examples underscore the importance of crafting prompts that require unique input to encourage meaningful learning experiences.

Table 4.2.1: Generalized versus Student-Engagement Assignments

Assignment	GPT Response	Analysis
Write a 50-word essay on a local event.	"In Pittsburgh, the annual Light Up Night heralds the holiday season, captivating attendees with spectacular fireworks, live music, and dazzling Christmas light displays. This beloved event fosters community spirit and boosts local businesses."	While acceptable for a Pittsburgh resident, this response would fail in Opelika, Alabama, highlighting the lack of assignment specificity. AI relies on probability to guess a location, often defaulting to widely known or generic events.

Table 4.2.2: Strategies to Encourage Student Engagement

Assignment	GPT Generic Response	Assignment with Student-Engaging	GPT Response to Additional Assignment	Analysis
Write a 50-word essay on a local event.	"In Pittsburgh, the annual Light Up Night heralds the holiday season, captivating attendees with spectacular fireworks, live music, and dazzling Christmas light displays. This beloved event fosters community spirit and boosts local businesses."	Write a 50-word essay on the history of **Carlisle, Ohio Railroad Days event.**	"The Carlisle, Ohio Railroad Days is an annual event celebrating the town's rich railroading heritage. Starting in the 1800s as a vital railroad hub, Carlisle honors this legacy with festivities, historical exhibits, and model train displays, drawing enthusiasts and community members to commemorate its pivotal role in regional development."	By specifying "Carlisle, Ohio Railroad Days," the assignment demands localized historical detail. Students must incorporate specific facts or class material to ensure alignment with the task, making AI-generated content more easily identifiable if unsupported by local knowledge.

In Table 4.2.1, the highlighted text in the " Assignment with Student-Engaging " column identifies the elements students must add to their responses. These additions compel students to incorporate personal knowledge, reflect on unique experiences, or directly reference classroom discussions. By embedding these requirements into assignments, educators ensure that students actively engage with the material, reducing the likelihood of AI dependency and fostering meaningful learning experiences.

4.3 Lack of Personal Voice

AI currently struggles to replicate the personal voice, unique perspective, and lived experience that define authentic student work. This limitation stems from the nature of AI systems, which generate responses using patterns from existing datasets rather than drawing on subjective memory or emotion. While AI can produce content that sounds personal, it cannot access actual experiences unless users explicitly provide them. For instance, if a student instructs the AI to include a detail like, "During a ski trip, I used leadership skills to help my group navigate a difficult situation," the tool can generate a polished reflection—but it cannot verify whether the event occurred or express how it felt. In this way, AI can simulate experience but not internalize it. Assignments that ask students to reflect on real moments or connect learning to their own development expose this gap. The result may be grammatically correct, but it often lacks the depth, inconsistency, or emotional tone that characterizes genuine introspection 【3】【4】.

Educators can use this limitation to design assignments that require students to demonstrate originality and personal insight. For example, a task asking students to describe how a specific classroom activity changed their understanding of teamwork demands context, emotion, and reflection. A student who actually participated must recall details and explain the learning process in their own words. In contrast, AI-generated responses often default to hypothetical summaries or abstract generalizations, omitting the subtle traits of lived experience. These types of assignments not only promote deeper student engagement but also make it easier for instructors to recognize when a response lacks authenticity. When specific, reflective detail is absent—or feels disconnected—AI dependency becomes more visible 【5】【6】.

The Table 4.3.1 below illustrates how the inclusion of personal experiences transforms a standard assignment into one that demands unique contributions, creating a meaningful distinction between authentic student work and AI-generated responses:

Table 4.3.1: Standard Assignments vs. Assignments with Personal Voice

Assignment Prompt	AI-Generated Response	Analysis
Write about a personal achievement that shaped your academic journey.	"Overcoming challenges in my studies taught me the importance of hard work and perseverance, helping me grow as a student."	The response is broad and lacks specific details, making it easily distinguishable as AI-generated. Authentic student responses would include unique events and personal reflections.
Reflect on how a class discussion influenced your view on teamwork.	"Teamwork is important because it allows people to collaborate and achieve common goals effectively."	The response is generic and does not reference the specific discussion or insights gained. A genuine response would include direct references to class content or team experiences.

The inability of AI to provide detailed and personalized responses presents an opportunity for educators to craft assignments that highlight the value of personal engagement. Tasks requiring students to articulate individual insights inherently demand input AI cannot replicate, bridging a critical gap in fostering authentic learning experiences. Table 4.3.2 illustrates how assignments designed to encourage personal voice compel students to integrate unique experiences, effectively reducing the reliance on AI-generated responses.:

Table 4.3.2: Personal Voice in Student-Engaging Assignments

Assignment Prompt	AI-Generated Response	Assignment with Student Engagement	GPT Response to Additional Assignment	Analysis
Write about a personal achievement that shaped your academic journey.	"Overcoming challenges in my studies taught me the importance of hard work and perseverance, helping me grow as a student."	Write a 60-word essay about how achieving a **4th-degree black belt,** a task that demanded discipline and hard work, influenced my academic journey. **This experience taught me the importance of knowledge and addressing gaps, which proved crucial as I pursued my Master's degree in Medical Simulation and a Doctorate in Education.**	"Achieving a 4th-degree black belt required immense discipline and hard work, instilling in me the importance of knowledge and gap identification. These skills translated seamlessly into my academic pursuits, culminating in a Master's degree in Medical Simulation and a Doctorate in Education, where the same principles of diligence and understanding were pivotal to my success."	The generic AI response lacks personal details and specificity, presenting a broad reflection that is easily identifiable as AI-generated. In contrast, the student-engaging assignment requires the integration of personal experience and individual reflection, forcing the student to provide unique insights into their academic journey. The detailed mention of earning a 4th-degree black belt and its impact on higher education demonstrates how personal voice and contextual engagement enhance the assignment.

Assignments emphasizing personal voice require students to incorporate unique experiences and specific details into their responses. The examples in the second table show how including individual achievements, such as earning a 4th-degree black belt, transforms a generic assignment into an engaging task. These adjustments challenge students to contribute their perspectives, fostering deeper learning and reducing reliance on AI-generated content. By prioritizing assignments that require personal reflection and context, educators can ensure authentic engagement and uphold the integrity of academic work.

4.4 Moving Beyond Surface-Level Analysis Through Metrics-Based Engagement

AI-generated responses often fall short when assignments require depth, critical engagement, or application of specific course materials. These outputs tend to provide surface-level summaries that, while coherent, lack analytical rigor and context. In areas like business education, where students must interpret regional policy, financial data, or strategic trends, surface-level AI responses are especially limiting. For example, a prompt asking for analysis of state-level tax incentives may elicit a general response like, "AI can help evaluate metrics such as employment growth and revenue change," without any localized insights or supporting evidence. Assignments like these reveal the gap between AI-generated content and student-developed analysis.

To overcome these limitations, students must engage with concrete, measurable data that adds specificity and relevance to their work. Assignments requiring students to research and apply key metrics—such as employment growth, business formation rates, or revenue trends—demand higher cognitive effort and deeper learning. By integrating these elements, students move beyond description and into interpretation. Tasks that call for comparisons between participating and non-participating businesses, or calculations of cost-per-job created, require critical thinking that cannot be fully automated.

Incorporating metrics into assignments does more than improve accuracy—it shifts ownership of the work to the student. AI tools cannot independently access or contextualize region-specific policy changes or localized economic data. As a result, students must seek out these details, evaluate them for credibility, and synthesize them into their arguments. This process ensures that learning remains active, not passive. Students are expected to demonstrate understanding through personalized analysis, reflection, and real-world data use.

Educators can reinforce this shift by structuring assignments around the integration of real-world data. A prompt might require students to calculate cost-per-job created based on actual budget data, or assess business survival rates in response to local tax policy. Rubrics can support this effort by emphasizing the importance of data integration, contextual understanding, and clear justification of conclusions. These practices limit AI's ability to deliver ready-made responses and increase the likelihood of authentic engagement.

By embedding detailed, metric-based criteria into assignments, educators reduce AI dependency while fostering independent thinking. Students are required to go beyond formulaic summaries and engage in critical analysis, bridging theoretical knowledge with practical application. Assignments framed this way support both academic rigor and professional relevance. They also teach students how to question, investigate, and reason—skills AI cannot replicate. This approach transforms learning from a transactional task into an intellectually active experience.

Table 4.4.1: Metrics-Based Assignments Across Educational Levels

Education Level	Assignment Prompt	Metric Required	Example Student Task
Elementary	Track the growth of a classroom plant over two weeks.	Daily height measurements	Record plant height each day and chart the changes using a line graph.
High School	Evaluate the success of your school's recycling program.	Monthly waste collection data	Compare pounds of recycled material per month and suggest improvements.
College	Analyze the impact of local tax incentives on small business growth.	Business formation and employment rate data	Use public data to assess policy outcomes and prepare a short policy brief.

4.5 Repetition of Patterns

AI often relies on repetitive phrasing and patterns embedded in its training data, resulting in responses that feel formulaic and predictable. These patterns emerge because AI generates outputs based on statistical probabilities, favoring common sequences over unique or novel constructions. Such tendencies make it difficult for AI to deviate from its learned structure, especially with general or open-ended prompts. Instructors familiar with AI tools can quickly recognize these repeated patterns, which often lack the originality, variation, and contextual adaptation found in authentic student work. Understanding these limitations helps educators design assignments that encourage deeper engagement and critical thinking, reducing overreliance on AI. AI's repetitive phrasing and predictable patterns make its responses formulaic and easy to spot, especially by instructors who know the nuances of AI-generated content.

AI-generated responses frequently rely on repetitive phrases and predictable language structures that reflect statistical probabilities learned during training. These patterns often lack the variation, nuance, and originality found in authentic student work. Educators who are familiar with AI outputs can quickly recognize these patterns, especially when evaluating generalized or vague assignments. To help educators identify such patterns, Table 4.5.1 highlights commonly overused words and phrases often found in AI-generated academic text. Understanding and anticipating these patterns allows instructors to design assignments that encourage originality and reduce the likelihood of AI-generated submissions passing as student work.

Table 4.5.1: Overused Words and Phrases in AI-Generated Text

Overused Words	Overused Phrases
Furthermore	It is important to note that...
Moreover	In conclusion...
Additionally,	This highlights the importance of...
Significant	Plays a crucial role in...
Important	One of the main factors is...
Impact	As a result of...
Result	Another aspect to consider is...
Key	The significance of this cannot be understated...
Benefit	Has had a major impact on...
Challenge	A key takeaway is...

AI often relies on repetitive phrasing and patterns embedded in its training data, which can result in responses that feel formulaic and predictable. These tendencies occur because predictive algorithms prioritize coherence over originality, favoring sequences that are statistically likely to appear in academic writing. Research has shown that AI systems like GPT frequently use a narrow range of transitional words and phrases—such as "furthermore" or "it is important to note that"—to maintain surface-level fluency【9】【10】. While these phrases may improve structural clarity, their overuse often signals a lack of depth, making AI-generated responses easier to recognize【11】. However, this pattern presents a challenge when evaluating the work of non-native English speakers, who may also rely on formulaic language as they strive to meet academic expectations in a second language. These students are sometimes flagged for AI dependency not because of misuse, but because their writing mirrors the polished, patterned tone commonly associated with AI tools. Educators who understand these distinctions can adjust their assignment design and feedback practices to better support all learners. Designing prompts that require precise, context-specific responses—grounded in classroom discussion or personal reflection—can help reduce reliance on AI tools while minimizing false positives for students still developing fluency【12】.

4.6 How AI Detectors Work: Understanding Patterns, Probability, Specificity

AI detectors operate by analyzing patterns in text and applying statistical models to assess the probability that a response was generated by an artificial intelligence system. The process begins by evaluating the structure of the writing, including sentence construction, transitional phrasing, and vocabulary usage. AI-generated responses often contain predictable patterns—repeated sentence structures, overused phrases, and generic terms—that reflect statistical norms drawn from training data.

Once patterns are detected, the system compares them against known samples of AI-generated and human-written content. This comparison is probabilistic, meaning the detector does not determine with certainty whether the text was AI-generated but calculates how likely it is based on similarity to previous examples. A high-probability score suggests that the response follows AI patterns closely. Common indicators include phrases such as 'It is important to note that' or words like 'significant' and 'impact,' which often appear in AI-generated responses due to their general applicability and polished tone.

Specificity also plays a critical role in this process. Responses that include context-specific details—such as references to classroom discussions, personalized examples, or localized data—are more likely to be recognized as human-written. These elements interrupt the pattern-recognition algorithm by introducing unique variables that AI tools typically do not generate on their own. In contrast, vague or general responses maintain a high degree of probability for AI generation because they closely match known AI output structures.

Instructors reviewing AI detector scores should understand that these tools rely heavily on the interplay between probability and specificity. High probability with low specificity suggests strong AI characteristics and little original student input. Low probability with high specificity often indicates authentic engagement, particularly when the text includes examples tied to recent classwork, individual insight, or community context. Understanding this process allows educators to make more informed judgments when interpreting AI detection results.

Table 4.6.1 below summarizes common phrases and words often flagged by AI detectors. These items are used as statistical markers to determine the likelihood of AI authorship and provide insight into how pattern-based scoring is applied.

Table 4.6.1: Common Linguistic Markers in AI Detection

Common Words	Common Phrases
Significant	It is important to note that...
Impact	In conclusion...
Result	This highlights the importance of...
Key	Plays a crucial role in...
Benefit	A key takeaway is...

4.7 Limitations of AI Detectors

Although AI detectors can be helpful in identifying AI-generated content, they have significant limitations. These tools primarily evaluate language patterns and statistical similarity to known AI outputs. As a result, they are susceptible to both false positives and false negatives, which can complicate their use in academic settings.

A false positive occurs when human-authored work is misclassified as AI-generated. For instance, a student using common transitional phrases such as 'moreover' or 'in conclusion' may be flagged incorrectly, even if the writing is original. Such errors can create unnecessary concern for both students and educators and may result in unwarranted accusations of misconduct.

A false negative happens when AI-generated content escapes detection. When students edit or personalize AI output by adding anecdotes, localized data, or contextual details, the final product can appear human-written. These submissions often blend AI structure with human insight, making them harder to identify through detection tools alone.

Educators should be aware that AI detectors operate without access to classroom context, assignment goals, or discussion-based insights. This lack of instructional awareness limits their ability to accurately judge whether student responses are AI-dependent or authentically produced. Understanding these weaknesses can help instructors interpret AI detection results more carefully and avoid placing too much trust in automated scoring tools.

False negatives represent another major limitation of AI detection tools, particularly when students engage AI strategically. There are two distinct types of false negatives. The first occurs when a student submits largely AI-generated content that is not detected due to minor rewording or structural changes. In this case, student engagement is minimal, and the integrity of the assignment is compromised. These are problematic because the learning outcomes are not met, even though the text bypasses detection.

The second type of false negative can reflect a more constructive scenario. Here, a student uses AI as a starting point but is required to add high levels of specificity—such as personal examples, localized data, or references to classroom content—to satisfy assignment requirements. As specificity increases, the probability of AI detection decreases because the output begins to resemble authentic human work. However, this is not inherently negative. In fact, when the student must modify AI-generated text through reflection and contextual detail, the process often leads to deeper learning and stronger engagement with the material.

Educators should distinguish between these two forms of false negatives. If the inclusion of student-driven specificity improves the quality of thought, supports critical thinking, and meets course objectives, then the presence of undetected AI elements is not cause for concern. The focus should remain on whether the student has actively participated in the learning process. In contrast, false negatives where no meaningful student input is evident require closer scrutiny, as they bypass the core intent of the assignment.

4.8 Shifting to a Classroom-Specific Assignment Model

AI tools such as ChatGPT and Claude generate responses based on existing training data. These systems do not possess awareness of classroom-specific details unless such information is explicitly included in the prompt. When students are provided with access to shared materials like slides, lecture notes, or recorded discussions, they can upload those resources along with a prompt to receive highly specific AI-generated responses. In such cases, the output may contain surface-level coherence yet give the impression of deep engagement. Educators must reconsider how much information and specificity is embedded in an assignment prompt, especially when materials provided can be paired with general tasks to produce plausible responses that may evade detection. The following examplse demonstrates a situation where a student pastes a general prompt into an AI tool without including course-specific content.

Assignment Prompt: *Using examples from our class discussion on educational technology, describe two ways to enhance student engagement.*

Claude AI Response*: "I apologize, but since I don't have access to your specific class discussions about educational technology, I can't reference those exact examples. However, I can suggest two evidence-based ways to enhance student engagement: Interactive polling and real-time feedback tools (like Kahoot or Mentimeter) boost participation by allowing students to respond anonymously and see collective results instantly. Project-based learning with technology integration lets students create digital artifacts (videos, presentations, websites) demonstrating their learning. This approach increases motivation by connecting abstract concepts to concrete outputs and giving students creative control."*

Analysis: The response includes general strategies but lacks connection to actual classroom content. The student would need to supplement the response with peer examples and instructor demonstrations discussed during class. Another case involves a student who includes lecture-specific information in the prompt.

Assignment Prompt: *Based on today's lecture on lesson planning, propose a strategy for adapting a math lesson to include diverse learners.*

ChatGPT Response*: "To effectively adapt a math lesson to include diverse learners, it's essential to consider various learning needs and styles. Here's a strategy that embraces this diversity: Differentiated Learning Objectives: Set tiered learning objectives that accommodate different levels of understanding and proficiency. Varied Instructional Methods: Use visual aids, auditory methods, kinesthetic activities, and technology integration to support diverse learners. Flexible Grouping: Rotate students through different groups for peer learning and mentorship. Assessment Variability: Offer multiple forms of assessment, such as tests, oral presentations, or practical demonstrations. Cultural Relevance: Incorporate* *examples reflecting students' diverse cultural backgrounds. Language Support: Provide glossaries and visual aids to support English language learners."*

Analysis: Although the response is comprehensive, it omits lecture-specific elements, such as manipulatives used in demonstrations or student-led strategies observed during the session. The student would need to incorporate those details to fulfill the intent of the assignment.

Both examples illustrate how specificity affects the relevance and detectability of AI-generated responses. Instructors must consider how shared course materials and detailed prompts can shape student engagement. When tasks require students to embed personal observations, cite live discussions, or refer to specific classroom events, AI cannot substitute that level of depth. Assignments should demand input that reflects the student's presence and participation in the course. Crafting prompts with this awareness helps ensure meaningful learning and reduces dependence on generative tools.

4.9 Encouraging Classroom Engagement Through Assignments

Assignments that explicitly reference classroom-specific materials prompt students to draw from their experiences and notes, ensuring deeper engagement. For instance, students in an education course might analyze a peer-reviewed journal article discussed in class and apply its findings to a teaching scenario. An AI-generated response would lack the insights from the class discussion, compelling students to engage with the material more actively to complete the assignment. By creating assignment prompts that require the use of classroom materials, instructors encourage active participation and discourage reliance on AI. For example, a task might ask students to evaluate a case study presented during class and relate it to theoretical concepts discussed in a prior lecture. These prompts require students to synthesize classroom knowledge, making AI-generated responses insufficient without personal engagement.

4.10 Educational Implications

Designing assignments that require the integration of classroom-specific content enhances learning by encouraging students to revisit and apply what they've learned. This approach reduces reliance on AI while promoting critical thinking and practical application.

Research indicates that assignments tied to course-specific materials lead to higher comprehension and retention, reinforcing the value of active learning 【13】.

Educators can create assignments that require students to use lecture materials, group discussions, or unique classroom examples. This strategy ensures that students develop a personalized understanding of the material and engage in higher-order thinking. Encouraging the application of class-specific concepts allows students to connect theoretical knowledge with practical scenarios, strengthening their academic growth and ability to apply what they've learned.

4.11 Recommendations for Educators

Educators can reduce student dependency on AI by designing assignments that require individualized input, localized content, or references to class-specific experiences. Rather than focusing solely on AI detection tools, instructors should prioritize instructional strategies that promote engagement, reflection, and intellectual ownership. Assignment prompts that ask for personal voice, metric-based analysis, or application of course materials limit AI's ability to generate plausible responses without student contribution. Tasks should ask students to interpret experiences, evaluate unique data sets, or reference peer discussions. These approaches allow instructors to assess understanding and reduce the effectiveness of generative tools when used without genuine engagement 【4】【5】【12】.

Rubrics should also be revised to assess depth of engagement, connection to classroom content, and the presence of reflective thought. Educators might include rubric categories such as "integration of personal insight," "use of class-specific materials," or "evidence of original interpretation." These components make it easier to distinguish between authentic responses and AI-generated text. Instructors can also adopt a transparency model that outlines expectations for AI use, allowing students to understand when AI support is appropriate and when original contributions are required 【3】【6】.

Support for educators should include professional development on AI literacy, assignment design, and feedback strategies that emphasize learning over detection. Faculty should be encouraged to test their prompts using AI tools to identify where specificity is needed and revise accordingly. Institutions may also provide shared repositories of AI-resistant assignments and rubrics tailored to different disciplines. By combining assignment design, transparent expectations, and instructional training, educators can create a learning environment where student thinking becomes visible and where AI, when used, contributes to rather than replaces academic development【4】【9】【12】.

4.12 Chapter 4 Summary

Chapter 4 examined how educators can detect and reduce AI dependency by identifying patterns of generalized, formulaic, or disconnected student responses. Indicators of AI use—such as vague phrasing, absence of personal voice, and failure to reference classroom content—highlighted the need for assignment redesign. Personal experience, localized data, and reflective insight emerged as central tools in promoting original thinking and distinguishing authentic student work. The chapter emphasized how vague prompts and surface-level tasks enable AI-generated responses that appear coherent but lack genuine engagement. Assignment examples and rubrics demonstrated how contextual input can increase specificity and student ownership of learning outcomes.

The chapter addressed the functionality and limitations of AI detection tools, noting their reliance on probability, pattern recognition, and linguistic markers. These tools provide useful insights but are insufficient without instructional context. Educators were encouraged to shift focus from detection to design by embedding metrics, reflection, and class-based analysis into assessments. This shift helps preserve academic integrity while promoting critical thinking. Chapter 4 underscored that reducing AI dependency requires proactive instructional design, emphasizing student learning through originality, specificity, and meaningful engagement with course content【4】【10】【12】【13】.

References

1. Bloom, B. S., Engelhart, M. D., Furst, E. J., Hill, W. H., & Krathwohl, D. R. (1956). *Taxonomy of educational objectives: The classification of educational goals. Handbook I: Cognitive domain*. Longmans, Green.
2. Dweck, C. S. (2006). *Mindset: The new psychology of success*. Random House.
3. Winkelmes, M.-A. (2013). Transparency in teaching: Faculty share data and improve students' learning. *Liberal Education, 99*(2), 48–55.
4. Zaphir, L., Lodge, J. M., Lisec, J., McGrath, D., & Khosravi, H. (2024). *How critically can an AI think? A framework for evaluating the quality of thinking of*

generative artificial intelligence (Version 2.0) [White paper]. arXiv. https://arxiv.org/abs/2406.14769

5. Vygotsky, L. S. (1978). *Mind in society: The development of higher psychological processes* (M. Cole, V. John-Steiner, S. Scribner, & E. Souberman, Eds.). Harvard University Press.
6. Knowles, M. S., Holton, E. F., & Swanson, R. A. (2015). *The adult learner: The definitive classic in adult education and human resource development* (8th ed.). Routledge.
7. Bransford, J. D., Brown, A. L., & Cocking, R. R. (Eds.). (2000). *How people learn: Brain, mind, experience, and school* (Expanded ed.). National Academy Press.
8. McCabe, D. L., & Treviño, L. K. (2002). Honesty and honor codes. *Change: The Magazine of Higher Learning, 34*(6), 32–38. https://doi.org/10.1080/00091380209605561
9. Brown, T., Mann, E., & Larsen, A. (2020). GPT-3: Language models are few-shot learners. *Advances in Neural Information Processing Systems, 33*, 1877–1901.
10. Clark, C., Watson, K., & Sullivan, R. (2021). Detecting AI-generated text: Identifying patterns in language models. *Journal of Artificial Intelligence Research, 71*, 235–250.
11. Johnson, D., & Hammer, P. (2019). The role of educational rubrics in a digital era. *Journal of Learning Sciences, 28*(4), 345–363.
12. Smith, T. (2022). Enhancing educational outcomes through AI-responsive assignments. *Educational Technology Research and Development, 70*(3), 413–428.
13. Nilson, L. B. (2016). *Teaching at its best: A research-based resource for college instructors* (4th ed.). Jossey-Bass.

Chapter 5 - The Role of Rubrics in Identifying AI Dependency

Rubrics are essential tools for establishing fair and transparent evaluation criteria, particularly in an academic environment shaped by AI tools like ChatGPT, Claude AI, and Bard. They help educators systematically assess originality, critical thinking, and student engagement with course material. However, two crucial elements must be considered when designing rubrics for AI-era education. First, AI operates on principles of probability and specificity, which means that a highly detailed rubric can inadvertently make assignments easier for students to complete using AI. Students may input the rubric into an AI tool, increasing the specificity and probability of generating a response that aligns closely with the evaluation criteria, often with minimal student input. These dynamic challenges traditional assumptions about rubric design and calls for a reevaluation of how rubrics can be structured to maintain academic integrity while promoting authentic learning. This chapter examines the dual role of rubrics in identifying AI dependency and offers strategies for balancing specificity with student engagement.

5.1 The Dual Role of Rubrics in the AI Era

Rubrics have long served as essential tools for clarifying expectations, guiding student performance, and promoting transparency in academic evaluation. In an environment shaped by artificial intelligence tools such as ChatGPT, Claude, and Bard, the role of rubrics has expanded. Rubrics now support instructors in detecting AI dependency while still encouraging authentic student engagement. A well-structured rubric can highlight the presence of originality, personal insight, and classroom-specific knowledge. However, rubrics that are overly detailed may also enable students to use AI tools more effectively by embedding the rubric into the prompt, producing content that aligns closely with evaluation criteria.

Table 5.1.1 demonstrates how minimal student input, such as inserting 'acting as me' and selecting a teaching strategy like 'flipped classroom,' enables generative AI tools to generate comprehensive responses. The examples included below reflect how AI can produce responses that align precisely with standard rubric expectations while requiring little critical thinking or personal reflection from the student.

Table 5.1.1: Examples Showing AI Dependency with Minimal Student Input

Scenario: The student added only "acting as me" and specified "flipped classroom" as the teaching strategy discussed in class. These minimal inputs allowed AI tools to generate comprehensive responses.

Assignment	Rubric Anchors	Student AI Prompt with Rubric
Write an analysis of a teaching strategy discussed in class.	Included classroom topic, included three examples, and followed the structure: introduction, examples, connection to theory, and conclusion.	**Acting as me**: Write an analysis of a **flipped classroom**, including three examples and connecting them to educational theories. Use the following structure: introduction, examples, connection to theory, and conclusion.

AI Tool Responses examples

Copilot

The flipped classroom strategy represents a shift in instructional design that emphasizes active learning and student-centered engagement. Rather than using class time for direct instruction, content is delivered outside of the classroom through videos or readings, allowing class sessions to focus on interactive learning experiences. This approach has been particularly effective in encouraging collaboration and critical thinking.

One example is the use of pre-recorded lectures, which students watch prior to attending class. This method allows for flexible learning and helps ensure that students arrive prepared to participate in activities. A second example includes in-class problem-solving exercises. These tasks allow students to apply concepts in real time, often working collaboratively to reach solutions. A third example involves structured peer discussions, which provide a platform for students to reflect on the content, ask questions, and build deeper understanding through dialogue.

The flipped classroom model aligns with several educational theories. Constructivism supports the idea that students build knowledge through active engagement and experience. Bloom's Taxonomy is also reflected in this approach, as students move beyond basic knowledge to apply, analyze, and evaluate material. Social Learning Theory reinforces the value of peer collaboration and interactive learning, which are core components of the flipped classroom model.

Overall, the flipped classroom strategy promotes deeper engagement by transforming the learning environment into one that values participation, collaboration, and critical thinking. It offers a model that not only improves academic performance but also fosters essential skills for future learning.

ChatGPT

The flipped classroom model has reshaped the traditional learning structure by moving content delivery outside the classroom and dedicating class time to active learning. This approach encourages deeper understanding by giving students the opportunity to interact with material in a more meaningful way during class sessions.

One example is found in physics courses, where students view instructional videos on topics like Newton's laws before class. During class, they engage in collaborative problem-solving that reinforces the material. Another example comes from medical education. Students watch videos on clinical techniques, such as suturing, and then practice those skills in hands-on lab sessions. A third example is used in business courses, where students review financial theories through readings or videos before analyzing real-world case studies in class with their peers.

These examples support several theoretical frameworks. Bloom's Taxonomy is evident in the emphasis on application and analysis during in-class activities. Vygotsky's Social Constructivism is also relevant, as students co-construct knowledge through collaboration and peer support. The flipped classroom fosters environments that align with these theories by prioritizing active participation and reflective practice.

In conclusion, the flipped classroom model provides a structure that enhances engagement and learning outcomes. By incorporating technology and prioritizing collaboration, it encourages students to take responsibility for their learning and develop skills essential to academic and professional success.

Claude Response

The flipped classroom represents a modern approach to education that reverses the traditional teaching model. In this structure, students are introduced to new content outside of class and then use class time to apply what they have learned through interactive activities. This method has helped me become a more engaged and independent learner.

In mathematics, for example, I watched instructional videos on calculus concepts before class, which allowed us to focus on solving complex problems together during class sessions. In literature courses, we read critical analyses and watched videos related to the texts before class. In class, we then discussed the material and participated in creative writing workshops. In science classes, pre-class materials covered theoretical content and lab safety protocols, enabling us to use class time efficiently for conducting experiments and analyzing results.

The flipped classroom approach aligns with several educational theories. Vygotsky's Social Learning Theory is evident in the emphasis on peer interaction and collaborative learning. Bloom's Taxonomy is addressed through higher-order thinking activities such as application and synthesis. Constructivist theory supports the value of experiential learning, which occurs through hands-on tasks and reflective discussion in class.

Overall, adopting the flipped classroom model has strengthened my ability to engage with course content, participate in meaningful dialogue, and develop real-world skills. This strategy has provided a foundation for deeper learning and greater academic success.

These examples reveal how minimal prompts can lead to full, structured responses aligned with academic rubrics. Although well-organized and theory-informed, these responses often lack meaningful personal engagement or classroom-specific insight. To understand the implications of this pattern, educators should consider the following points:

1. Reduced Student Effort: Minimal input limits critical thinking, as AI fills in details using generalized patterns.
2. AI Limitations: While polished, the responses lack references to peer dialogue or instructional nuances unique to the class.
3. Encouraging Student Engagement: Assignments should require personal or course-specific content, ensuring student participation.
4. Implications for Rubric Design: Rubrics should reward reflection, context, and individualized insight, discouraging generic responses.

5.2 How Assignment and Rubric Changes Foster Student Engagement

Revised assignments and rubrics foster student engagement by shifting the focus from surface-level responses to an in-depth exploration of material that demands active participation and critical thinking. By integrating specific criteria, the revised rubric compels students to engage with the assignment meaningfully, ensuring they demonstrate not only knowledge but also the ability to apply it within a contextual framework. This approach transforms assignments into exercises of personal and academic growth, where students develop a deeper connection to the subject matter.

5.3 Assignment Changes That Promote Engagement

The revised assignment requires students to analyze a teaching strategy discussed in class, incorporating three specific examples and connecting them to course theories and peer insights. Unlike generic prompts, this task demands that students:

1. **Reflect on Classroom Discussions:** Students must recall and analyze specific conversations or activities from class sessions. This step requires active listening, note-taking, and an understanding of how these discussions connect to broader educational themes.
2. **Identify Relevant Examples:** Selecting three contextually relevant examples forces students to critically evaluate the material and identify moments from class that align with the assignment's objectives. This process ensures that students are not merely summarizing but actively synthesizing information.

3. **Apply Theoretical Frameworks:** Connecting classroom examples to course theories requires students to bridge abstract concepts and practical applications. This step deepens their understanding of how theoretical principles operate in real-world scenarios.
4. **Integrate Peer Insights:** Incorporating perspectives from classmates encourages collaboration and highlights the value of diverse viewpoints, fostering a more comprehensive and nuanced analysis.
5. **Demonstrate Critical Thinking:** The assignment requires students to not only describe examples but also critically evaluate the effectiveness of the teaching strategy in relation to course theories and peer feedback.

5.4 Rubric Anchors That Drive Engagement

Rubric anchors guide the evaluation of student work and serve as a framework for encouraging critical thinking and course engagement. In the context of AI-enhanced learning, anchors that are too specific can inadvertently increase the probability that AI-generated content aligns with grading expectations. While specificity improves clarity, over-specification—such as listing exact diagnoses or steps—can enable students to prompt AI tools with a grading key. Experienced instructional designers balance this risk by structuring rubric anchors that emphasize reasoning, course integration, and personal insight, rather than task completion alone.

The following table compares rubric anchors that promote deeper engagement with those that may support surface-level, AI-driven responses, along with rationale for each distinction.

Table 5.4.1: Comparison of Rubric Anchors – Engagement vs. AI-Predictable Responses

Surface-Level Anchor (AI-Predictable)	Engagement-Driven Anchor	Rationale
Student explained differential diagnoses of CHF and COPD.	Student explained how two differential diagnoses were ruled out using classroom discussion or patient-specific symptoms.	Explicit diagnosis names increase the probability of AI generating a targeted response. Emphasizing reasoning and course content limits direct AI use and promotes independent analysis.
Student referenced Bloom's Taxonomy and Vygotsky's theory.	Student connected the selected strategy to theoretical concepts discussed in class and explained how these theories informed student-centered practices.	Naming specific theories allows AI to generate structured comparisons. Asking for applied connections to course concepts increases engagement and reduces AI predictability.
Student listed three examples of the flipped classroom model.	Student provided three examples of the flipped classroom approach grounded in peer demonstrations or course media.	Listing examples encourages AI summarization. Requiring contextual links to course sources or peer contributions encourages student participation.
Student reflected on the Week 3 activity and summarized communication style.	Student analyzed their own communication during Week 3 feedback rounds, including strengths, areas for growth, and reference to rubric criteria.	Surface-level summaries can be AI-assisted. Anchors requiring personal reflection, rubric language, and self-evaluation require authentic engagement.
Student described two instructional strategies covered in lecture.	Student selected two instructional strategies discussed in lecture and	Describing is generic and predictable. Asking students to link strategies to course

	explained how they were applied in classroom activities or simulations.	activities requires synthesis and limits AI reuse.

Effective rubric anchors create a barrier between surface-level completion and genuine learning. Anchors that invite interpretation, classroom-specific references, and personal accountability elevate expectations and make AI-generated completion less viable. Instructional designers must consider both the probability of AI alignment and the level of specificity required to assess original student thinking. The revised rubric includes specific anchors that direct students toward active participation and thoughtful analysis. Each anchor reinforces a layer of complexity that encourages unique, grounded, and personally meaningful responses.

1. References to Classroom Discussions and Peer Insights:

   Student integrates insights shared by peers or instructors during class discussions. This ensures that the response reflects presence, attentiveness, and the application of real-time classroom dialogue.

2. Detailed and Contextually Relevant Examples:

   Student identifies and analyzes three specific examples relevant to course objectives. Examples should draw from course materials, classroom experiences, or personal observations to discourage generalized or templated answers.

3. Connections to Course Theories:

   Student links the assignment topic to key theoretical frameworks introduced in class. Rather than listing the names of theories, the student demonstrates how those concepts shaped instructional strategies or decisions.

4. Critical Thinking and Personal Engagement:

   Student critiques the effectiveness of a strategy or reflects on its relevance to their own development. Evaluation must include reasoning and evidence of individualized understanding.

5. Structured Format:

Student presents their work using a consistent structure—introduction, examples, theoretical connection, and conclusion—to ensure clarity, cohesion, and progression of ideas.

5.5 What Students Must Add to Meet the Assignment Requirements

To properly meet the revised assignment requirements, students must add the following:

1. **Specific Examples:** Students need to provide detailed instances from classroom discussions, describing what was said, by whom, and how it relates to the teaching strategy being analyzed.
2. **Personal Observations:** Including personal reflections and interpretations demonstrates their engagement and adds a unique perspective.
3. **Course Theory Integration:** Students must explicitly connect the examples to theoretical frameworks discussed in the course, showing their ability to apply abstract ideas.
4. **Peer Perspectives:** Incorporating insights from classmates requires students to acknowledge and analyze different viewpoints, adding depth to their response.
5. **Critical Evaluation:** Students must assess the effectiveness of the teaching strategy, offering insights into its strengths, weaknesses, and potential improvements.
6. **Clear Organization:** A structured response with logical flow ensures that their analysis is easy to follow and professionally presented.

By addressing these requirements, students move beyond rote memorization or generic summaries, engaging deeply with the course material. This process enhances their critical thinking, fosters a sense of ownership over their learning, and develops their ability to connect theory to practice effectively.

5.5.1 Example of Revised Rubric and Assignment

Assignment	Rubric Anchors	Student AI Prompt with Rubric Input
Write an analysis of a teaching strategy discussed in class, providing three specific examples from our classroom discussions and connecting them to course theories and peer insights.	- Includes references to specific classroom discussions or peer insights. - Provides three detailed examples of the teaching strategy. - Follows the structure: introduction, examples, connection to theory, and conclusion. - Demonstrates critical thinking and personal engagement with the material.	Acting as me: Write an analysis of a teaching strategy discussed in class, providing three specific examples from our classroom discussions and connecting them to course theories and peer insights. Includes references to specific classroom discussions or peer insights. - Provides three detailed examples of the teaching strategy. - Follows the structure: introduction, examples, connection to theory, and conclusion. - Demonstrates critical thinking and personal engagement with the material.

Claude AI Response

> I apologize, but I should first clarify: Without having been present in your specific classroom discussions, I wouldn't be able to provide authentic examples from your class or reference specific peer insights. However, I'd be happy to help you structure your analysis if you could share:
>
> - Which teaching strategy you'd like to analyze
> - Some examples from your classroom discussions
>
> The course theories you've covered
>
> This way, I can help you develop a well-structured analysis that genuinely reflects your classroom experience and connects to the specific theories you've studied. Would you be willing to share these details?

Claude AI's response highlights its inability to access classroom-specific discussions or peer insights. This limitation emphasizes the importance of personalized input from students. By requiring students to reference detailed examples from their classroom experiences, the revised rubric ensures that students engage deeply with the material. AI becomes an assistant in structuring and organizing the response, but it cannot substitute the student's active contribution. For example, a prompt that asks, *"Describe two teaching strategies discussed in last week's class and explain how your group applied them during the microteaching session,"* demands knowledge that is only available to students who were present and participating. If a student attempts to outsource this task to AI, the tool can only provide general strategies like differentiated instruction or flipped classrooms—but it cannot describe specific strategies demonstrated in the actual session, or explain how the student's group used them.

Consider a hypothetical case in which a student uses AI to respond to an assignment prompt asking them to evaluate peer feedback given during a classroom roleplay. The AI may generate plausible comments like, *"My peer emphasized the importance of clear instructions and maintaining eye contact."* While coherent, these responses are easily identified as generic. In contrast, a student who was engaged might write, *"During our roleplay, Emma suggested I use fewer instructions at once because the pacing overwhelmed our learners. This feedback helped me revise my approach in our second session."* That level of personal detail demonstrates engagement and learning—something AI cannot generate without being fed the full experience. Requiring this kind of specificity pushes students to participate, reflect, and connect learning to their own process, effectively reducing overreliance on AI.

5.6 Full Student Engagement: Leveraging AI as a Tool, Not a Replacement

In the evolving landscape of AI-assisted education, the distinction between full student engagement and passive AI reliance is critical. Students who actively engage in their assignments by integrating detailed classroom content and personal insights transform AI from a standalone solution into a collaborative tool. The following example illustrates how a student can effectively prompt AI for assistance while maintaining ownership of their academic work.

Prompt for Full Student Engagement

Student's AI Prompt: *Acting as me: Write an analysis of a teaching strategy discussed in class, providing three specific examples from our classroom discussions and connecting them to course theories and peer insights. Include references to specific classroom discussions or peer insights. Provide three detailed examples of the teaching strategy. Follow the structure: introduction, examples, connection to theory, and conclusion. Demonstrate critical thinking and personal engagement with the material. Use the flipped classroom strategy, referencing the following classroom discussions:*

1. **Pathophysiology of Cardiac Conditions**
   Students watched a pre-recorded lecture on the pathophysiology of myocardial infarction before class. During in-class sessions, they collaborated in small groups to analyze clinical scenarios, interpret ECGs, and determine appropriate interventions with instructor guidance.
2. **Pharmacology of Antibiotics**
   Pre-class materials included a video explaining antibiotic mechanisms and resistance. In class, students engaged in case-based discussions to select appropriate antibiotics for specific infections, applying pre-learned concepts to real-world situations.

3. **Clinical Skills Practice**

Students reviewed pre-class videos on physical examination techniques for abdominal pain. In-class activities involved practicing the techniques on standardized patients under faculty supervision, focusing on skill refinement and diagnostic reasoning.

Course Theories to Incorporate:

1. **Pathophysiology of Cardiac Conditions:** Bloom's Taxonomy and Constructivist Learning Theory.
2. **Pharmacology of Antibiotics:** Situated Learning Theory and Kolb's Experiential Learning Theory.
3. **Clinical Skills Practice:** Vygotsky's Social Constructivism and Experiential Learning Theory.

Analysis of Full Student Engagement

The prompt showcases high student engagement as it demands extensive preparation and input before AI can be utilized effectively. Here's how this approach demonstrates robust student involvement:

Integration of Specific Content:

1. The student must reference specific classroom activities and discussions, such as group ECG analysis, antibiotic selection cases, and clinical skills practice. These details require students to reflect on their learning experiences, ensuring deep interaction with the material.

Connection to Course Theories:

2. The student incorporates educational theories discussed during the course, such as Bloom's Taxonomy and Vygotsky's Social Constructivism. This connection reinforces the student's understanding of theoretical frameworks and their application.

Demonstration of Critical Thinking:

3. By asking the AI to integrate classroom discussions with course theories, the student demonstrates analytical skills and contextual understanding. This approach ensures that the output aligns with the course's unique requirements.

Ownership of Content:

4. The student's detailed prompt indicates a strong grasp of the material, positioning AI as a secondary tool for organizing and enhancing their work. The personalized input ensures that the response reflects the student's voice and perspective.

Comparison of Engagement Levels

Engagement Level	Student Input	AI Role
Minimal Engagement	"Acting as me: Write an analysis of a flipped classroom."	AI generates a generic response, lacking depth, classroom references, or personal insights.
Full Engagement	Detailed prompt specifying classroom discussions, peer insights, and relevant theories.	AI assists in structuring and refining the response but relies on student-provided details to ensure accuracy.

5.7 Encouraging Meaningful Engagement: Strategies for Rubric and Assignment Design

The challenge for educators is creating rubrics and assignments that encourage students to use AI as a supportive tool while ensuring active engagement in the learning process. Overreliance on AI often results from prompts and rubrics that unintentionally allow AI to generate responses with minimal student input. By thoughtfully structuring rubrics and assignments, educators can guide students toward deeper engagement and critical thinking. Below are strategies for designing assignments and rubrics that encourage authentic student contributions:

1. Reference Specific Classroom Content

Assignments should require students to incorporate unique classroom content that AI cannot access or replicate.

- **Examples of Classroom-Specific Tasks:**
    - Summarizing peer-led presentations or discussions.
    - Analyzing case studies discussed in class.

- Referencing lecture-specific analogies or scenarios provided by the instructor.

- **AI Limitation:** AI lacks access to proprietary classroom materials, discussions, or spontaneous interactions. Assignments based on these unique elements ensure students must rely on attendance and participation to succeed.

2. Reflect on Personal Insights and Experiences

Prompts should encourage students to include personal experiences and perspectives unique to their learning journey.

- **Examples of Reflective Tasks:**
  - Analyzing how a group project or peer feedback session improved their understanding of a subject.
  - Describing how their background or prior experiences shaped their perspective on a class topic.
  - Journaling about challenges they overcame during a classroom activity.
- **AI Limitation:** AI cannot replicate a student's personal journey, emotions, or individual growth, making this type of engagement distinctly human.

3. Analyze Theories Through Classroom Context

Students should integrate theories or frameworks with examples from classroom discussions and readings. This approach deepens their understanding and ensures originality.

- **Examples of Analytical Tasks:**
  - Applying a specific theory to an example from a class activity.
  - Comparing different theories discussed during lectures and evaluating their relevance to a case study.
  - Creating diagrams or flowcharts to visually represent the relationship between classroom theories and real-world applications.
- **AI Limitation:** While AI can summarize theories, it cannot contextualize them with insights drawn from classroom discourse or unique instructional materials.

4. Incorporate Group Assignments and Collaborative Work

Group-based activities require interpersonal interaction, negotiation, and shared contributions, which AI cannot replicate. For instance:

- **Examples of Collaborative Assignments:**
  - Developing a group presentation based on a class project.
  - Participating in debates or peer reviews during class.
  - Building a shared portfolio or collective research project.
- **AI Limitation:** Collaboration requires human interaction, adaptability, and the ability to synthesize diverse perspectives. AI cannot mimic the dynamic nature of human teamwork or contribute meaningfully to collective decision-making processes.

5. Require Multimodal Assignments

Assignments involving diverse formats or outputs, such as visual aids or oral presentations, compel students to engage actively. For instance:

- **Examples of Multimodal Assignments:**
  - Creating a poster or infographic summarizing a classroom experiment.
  - Delivering a live presentation supported by slides or props.
  - Producing a video reflection or demonstration of a skill learned in class.
- **AI Limitation:** AI cannot deliver live presentations or create physical objects. These tasks demand human creativity, communication skills, and direct participation.

Rubrics and assignments that encourage referencing classroom content, reflecting on personal insights, analyzing theories through class discussions, collaborating with peers, and producing multimodal outputs push students beyond passive learning. By carefully designing these elements, educators can ensure that AI serves as a valuable support tool without replacing the active engagement and critical thinking essential for meaningful education.

5.8 Designing Rubrics That Detect AI Dependency

To counteract the potential misuse of rubrics, educators must design rubrics that encourage authentic student engagement while identifying signs of AI dependency. The core elements of an AI-detection rubric should include:

5.8.1 Core Elements of an AI-Detection Rubric with Examples

Rubric Element	Description	Example Assignment	Analysis
Originality	Assess whether the response reflects unique insights, personal experiences, or classroom connections.	Reflect on how a class discussion influenced your approach to classroom management.	Authentic responses would include specific references to class discussions. AI-generated responses often omit these details, making dependency easier to detect.
Depth of Analysis	Evaluate the extent to which the response provides detailed reasoning, critical evaluation, or contextual understanding.	Critique the challenges of integrating project-based learning into high school classrooms.	Superficial analysis or generic strategies reveal AI reliance. Authentic responses would include nuanced arguments based on classroom discussions.
Alignment with Content	Ensure the response integrates specific references to lectures, readings, or discussions.	Based on today's lecture, analyze the use of scaffolding in math instruction.	Responses that fail to reference specific examples or techniques shared during class demonstrate AI dependency.
Voice and Personalization	Look for evidence of the student's individual perspective and reflective thinking.	Write about a personal teaching experience that shaped your classroom philosophy.	AI-generated responses are likely to sound detached. Authentic student work incorporates personal

Rubric Element	Description	Example Assignment	Analysis
			anecdotes, providing a unique and engaging perspective.
Cohesion and Logical Flow	Examine whether the response demonstrates a coherent structure and logical progression of ideas.	Propose a comprehensive strategy for enhancing group collaboration in classroom settings.	AI-generated responses may lack seamless transitions or logical progression, making them identifiable as formulaic outputs.

5.9 Chapter 5 Summary

Chapter 5 explores the evolving role of rubrics in an AI-influenced educational environment. Traditionally used to guide student work and ensure consistent grading, rubrics now serve a dual purpose—assessing both student learning and AI dependency. The chapter outlines how AI tools can exploit overly specific rubrics by using rubric input to generate polished responses that align closely with evaluation criteria. These outputs, while technically correct, often lack critical thinking, personal insight, or contextual relevance【1】【2】.

To address these concerns, the chapter offers strategies for balancing clarity with intellectual challenge. Educators are encouraged to revise assignments and rubrics to include open-ended prompts, references to class discussions, and opportunities for personal reflection. Examples show how minimal student input, such as specifying a teaching strategy and copying rubric instructions, can lead to fully structured AI-generated responses【1】【3】. In contrast, more rigorous tasks demand that students incorporate personal analysis or contextual evidence, reducing the likelihood of AI-generated work passing as original【4】.

A growing risk in rubric design is the problem of overthinking—where instructors attempt to preempt AI misuse by including so much detail that the rubric itself becomes a blueprint for a successful AI prompt. Students who recognize this may bypass engagement by pasting the rubric into AI tools like Claude or ChatGPT, effectively receiving a full response with little to no effort 【2】【4】. Overthinking in rubric design may come from a desire to maintain clarity, but excessive specificity can result in unintended shortcuts. Educators must find balance, ensuring rubrics support authentic engagement while preserving academic integrity. The goal is to encourage student thinking, not automate it.

References

1. Brown, T., et al. (2020). GPT-3: Language Models Are Few-Shot Learners. *Advances in Neural Information Processing Systems*, 33, 1877-1901.
2. Clark, C., et al. (2021). Detecting AI-Generated Text: Identifying Patterns in Language Models. *Journal of Artificial Intelligence Research*, 71, 235–250.
3. Johnson, D., & Hammer, P. (2019). The Role of Educational Rubrics in a Digital Era. *Journal of Learning Sciences*, 28(4), 345–363.
4. Smith, T. (2022). Enhancing Educational Outcomes through AI-Responsive Assignments. *Educational Technology Research and Development*, 70(3), 413–428.

Chapter 6: Practical Applications Across Educational Levels

Integrating artificial intelligence into education requires more than awareness—it requires application. While many conversations about AI in learning focus on policy, risk, or theoretical impact, the real shift happens at the level of classroom practice. The question for most educators is not whether students will use AI, but how to design environments where learning still occurs when they do. In medical education, we face this challenge daily. Whether training a student to write a SOAP note, diagnose a case, or practice communication skills with a simulated patient, the focus is always the same: meaningful engagement. In my own experience designing simulations, the most effective activities were not the ones that asked students to repeat information—but those that required them to apply what they knew to a context that changed in real time. AI creates a similar demand: if educators want students to think, they have to assign work that can't be completed with a copy-paste solution.

Across educational levels, assignments can be adapted to limit AI dependency without needing restrictive policies. In medical education, I've built case studies that evolve based on a student's choices. For example, when students are asked to explain why they chose a particular intervention during a clinical simulation, they must draw on lectures, clinical guidelines, and their own judgment—not just produce a technically correct answer. The same principle can apply in K–12 or undergraduate settings. Whether students are describing a local public health initiative, evaluating a peer-led teaching strategy, or explaining how their group reached a decision during a lab activity, tasks that require personal reasoning, course-specific input, or live reflection become difficult for AI to replicate. This chapter outlines how to implement those practices across contexts using concrete, adaptable examples.

6.1 Leveraging AI for Enhanced Learning and Student Engagement

Artificial intelligence can enhance student engagement when used to support—not replace—thinking. Across educational levels, AI has potential to help students organize ideas, test knowledge, and reflect on their learning process. However, the value of AI depends entirely on how the task is designed. If students can copy and paste a question into AI and receive a complete answer, meaningful engagement is lost. When the assignment instead requires students to make choices, add context, or critique what the AI returns, the tool becomes a support—not a shortcut. In medical simulation, this principle is well established. The student must make a decision before the simulation responds. That demand for action is what keeps learners engaged—and what allows them to suspend disbelief long enough to treat a manikin like a real patient. Assignments should function the same way: they must require student input before progress can occur.

In simulation design, manikin-based scenarios include decision points that require student participation. Blood pressure readings are not provided until the student applies the cuff and explicitly asks for them. Lab results are not delivered unless the student specifies the exact test. General requests such as "run some labs" receive no response. Learners must name individual diagnostics—like *"order a CBC and lactate"*—to receive the next piece of information. The scenario only progresses when the student acts, reinforcing the habit of clinical reasoning through intentional engagement.

Assignments can follow a similar structure. For example, asking, *"What are the lab findings for a patient with sepsis?"* allows AI to generate a complete answer without the student engaging in the decision-making process. A more effective prompt would be, *"Identify three lab tests you would prioritize for a 70-year-old patient with hypotension and confusion. Justify each choice based on symptoms and clinical need."* The second version requires the student to provide context before AI can contribute. Any AI use becomes secondary to student judgment.

Similar models apply outside of medicine. In an English class, a student may use AI to explore themes from a novel but must select one and write an analysis supported by class discussions. In a history course, AI might be used to generate a timeline, but students are expected to annotate events using content from lectures and primary sources. In each case, student thinking drives the task—not the AI's response.

Whether in simulation or academic coursework, learning depends on participation. A manikin only becomes a patient when the student treats it like one. An assignment becomes a thinking exercise only when it requires interpretation, specificity, and reflection. Building tasks that expect students to act before they receive support—whether from AI or a simulation—ensures the learning remains theirs.

6.2 Grading and Feedback Using AI

Artificial intelligence can support the feedback process by helping students identify strengths and weaknesses in their work before they submit it. Many AI tools allow students to input a rubric alongside a draft to receive comments aligned with assignment expectations. Rather than waiting for instructor feedback after submission, students receive structured guidance that can help them revise early in the process. For many learners—especially those still developing writing confidence or content knowledge—this opportunity improves clarity and supports reflection before a grade is assigned.

In medical education, students often use rubrics to guide their documentation of clinical reasoning. When drafting SOAP notes or simulation reflections, students can request AI feedback on organization, tone, or clarity. For example, a student may paste their draft and rubric into an AI prompt and ask for suggestions to improve clinical justification in the assessment section. The AI might recommend stronger connections between objective findings and diagnostic reasoning, allowing the student to revise before submission.

Students in other disciplines can use AI in similar ways. A high school student writing a persuasive essay can ask whether the thesis is clear or whether evidence is well-distributed across paragraphs. A college student working on a case study might ask for help identifying whether the analysis section includes both observation and interpretation. The feedback they receive does not replace instruction but gives students an early diagnostic tool—similar to how a checklist might help them verify procedural steps before turning in a lab report or research paper.

Educators can design assignments that include this step by requiring students to submit both their original draft and the AI feedback they received. That practice not only encourages students to use the tool for revision rather than generation, but also provides insight into how students are learning to self-correct. Instructors can review both documents to focus grading on higher-order issues and to verify whether revisions reflect meaningful engagement. In this way, AI becomes a feedback scaffold rather than a substitute for thinking.

Incorporating AI into the feedback cycle prepares students to approach revision with more intention and builds habits of independent review. Assignments structured in this way encourage students to take responsibility for identifying and correcting their own mistakes—an essential component of professional and academic growth.

6.3 Ensuring Assignment Completion

Artificial intelligence can help students confirm they have met all assignment requirements before submission. When paired with a copy of the prompt or rubric, tools such as ChatGPT or Claude can review a student's draft and identify missing components, vague language, or incomplete reasoning. Students benefit from structured, real-time feedback that often highlights areas needing improvement—before an instructor ever sees the work.

In medical education, this practice has practical applications. A student preparing a SOAP note can input the rubric along with their draft and request feedback on each required section: subjective, objective, assessment, and plan. If the "assessment" lacks clear clinical linkage to objective findings, the AI can prompt the student to improve reasoning or restructure the section. Rather than waiting for instructor correction, students can identify and fix those gaps independently.

Educators who adopt this technique may consider asking students to submit both the original AI feedback and the revised assignment. That approach mirrors what many math courses already require—students must show the steps, not just the answer. In the same way, students can submit a screenshot or document showing the AI's evaluation and their revised response. Doing so creates a trail of thinking, promotes metacognitive reflection, and reduces the burden on the instructor to flag missing elements after the fact. For the student, the AI becomes a kind of tutor: one that helps them practice identifying errors, applying corrections, and gaining confidence in the material. For the educator, the submitted AI output provides a clear view of what students improved independently—making grading faster and more focused on deeper learning.

The same method can be used by educators to streamline grading. By entering the assignment and rubric into an AI tool, instructors can receive a preliminary evaluation that highlights strengths and weaknesses in each submission. Reviewing the high and low points of an AI-generated summary for each student paper can support consistency and reduce fatigue during long grading sessions. Educators remain the final decision-makers, but the AI can serve as a first-pass reviewer—offering structure, reducing cognitive overload, and helping identify patterns across submissions. Instructors using this method should always follow institutional policies on AI use and ensure any AI-based evaluation remains compliant with privacy, transparency, and grading standards. [1]

Designing assignments to encourage AI-supported self-review and instructor-assistive feedback supports both efficiency and learning. When implemented responsibly, this approach builds student agency, reduces grading time, and enhances consistency without compromising instructional quality.

6.4 Creating Study Materials from Content

Artificial intelligence allows students to turn static learning materials into active review tools. One of the most effective applications is the ability to convert lecture content, textbook excerpts, or presentation slides into personalized study questions. Students can input course material into AI tools and request multiple-choice, true-or-false, or short-answer questions that align with their study needs. By engaging with content through questioning, students move beyond passive reading and toward more active processing.

In a college-level psychology course, for example, a student studying developmental theories might upload a section of lecture notes and ask the AI to create five multiple-choice questions focused on Erikson's stages of development. The student can then quiz themselves or a classmate, using the responses to identify areas of confusion or reinforcement. A high school history student might upload a chapter on the American Revolution and request a mix of true-or-false and short-answer questions to test their understanding of major events, figures, and causes.

Students who benefit from alternative approaches to studying often find this method helpful. Learners with dyslexia may struggle with dense text, but can more easily engage with bite-sized, structured prompts. Requesting AI to generate questions not only reduces reading fatigue but also promotes retrieval practice and critical thinking. Visual and auditory learners can adapt these outputs for flashcards, oral quizzes, or interactive formats to reinforce key concepts.

Other creative uses of AI in study design include thematic organization of lecture content. A student can ask AI to summarize a multi-topic lecture by identifying central themes and organizing supporting details accordingly. For instance, an economics student might request a thematic breakdown of content related to fiscal policy, international trade, and market regulation across several weeks of lecture slides. AI can help condense content into manageable clusters that reinforce concept mapping and long-term understanding.

AI can also be used to gamify review sessions. Students may input material and ask the AI to design a quiz show, flashcard deck, or interactive game format. For example, a group of students preparing for a science test might ask AI to convert key terms and definitions into a Jeopardy-style review game they can use in a group setting. These game-based formats enhance engagement, especially for learners who respond well to competition, collaboration, or movement-based review.

Educators can intentionally incorporate this approach into assignments. For example, instructors might require students to generate five questions based on weekly readings, explain the purpose of each question, and submit both the questions and answers. Structuring tasks in this way encourages students to prioritize key material and practice framing information in ways that support comprehension. AI serves as a bridge—not a replacement—between content delivery and meaningful review.

Research supports the connection between active study methods and improved learning outcomes, particularly for students with learning differences【2】. Using AI in this context supports personalized review while also building study habits that promote long-term retention and academic confidence. [2].

6.5 Supporting Critical Engagement

Artificial intelligence can support critical thinking when students are asked to evaluate, refine, or challenge AI-generated content. Rather than viewing AI as a replacement for originality, instructors can incorporate it into assignments that require deeper reasoning and judgment. When students are expected to interact with what the AI produces—rather than simply submit it—they begin to develop critical engagement skills that strengthen their ability to analyze, synthesize, and defend ideas.

In a communication course, students might prompt AI to draft a response to a workplace conflict scenario. After reviewing the result, they could be asked to assess whether the tone was appropriate, whether the message aligned with organizational values, and how different audiences might perceive the content. Students begin to recognize the importance of nuance in professional communication, understanding that effective messaging depends not only on content but also on delivery and emotional awareness.

A similar approach works well in the sciences. A biology student could enter lab data into an AI prompt and ask for a potential hypothesis. Once the response is received, the student is asked to critique its scientific validity based on course readings or lab results. Engaging with the AI output becomes a structured point of reflection, where students identify errors, question assumptions, and apply foundational concepts.

In a business course, students might use AI to brainstorm names for a new product or service. After reviewing several AI-generated options, each student is asked to select one and justify their choice based on target audience, brand identity, and marketing appeal. During a class presentation, the student explains not only the selection but also why other options were rejected. In doing so, students engage in branding strategy, persuasive reasoning, and audience analysis—all framed through critique and decision-making rather than passive generation.

Assignments that require students to question, revise, or respond to AI-generated work build essential habits of intellectual curiosity and responsibility. Students begin to see that learning is not about receiving correct answers but about interpreting information, applying knowledge, and defending decisions. When AI is used to prompt analysis instead of replace it, students remain at the center of the learning process. [3].

6.6 Ethical Integration of AI Tools

Educators play a critical role in guiding students toward responsible use of AI in their academic journey. Promoting ethical AI practices ensures students rely on these tools for support rather than as a substitute for their effort. Transparency about the capabilities and limitations of AI fosters trust and encourages informed decision-making among students. Teaching students to integrate AI thoughtfully aligns with active learning principles, emphasizing their role in the educational process. Preparing students for a technology-driven world requires practical skills and a strong foundation in ethical technology use. Research supports the importance of developing digital literacy as a fundamental component of modern education [4].

Peer-reviewed evidence supports the integration of AI as a tool to promote self-directed learning and improve academic outcomes. AI tools encourage students to take ownership of their educational experience while accommodating diverse learning styles. Educators who thoughtfully implement these tools create an inclusive environment that prepares students for future challenges in academia and beyond. [4]

6.7 Chapter 6 Summary

Chapter 6 offers a practical exploration of how artificial intelligence can be integrated into classroom learning across multiple educational levels. Rather than focusing on policy or abstract concerns, the chapter centers on instructional strategies that enhance engagement, support feedback, and foster critical thinking. Educators are encouraged to design assignments that require student input, reflection, and contextual awareness—features that limit AI's ability to generate complete responses. Through medical simulation and case-based examples, the chapter demonstrates how student action must drive instructional outcomes. Whether in secondary or higher education, the effectiveness of AI integration depends on how well tasks maintain student agency.

Assignments can also be structured to support revision, personalization, and deeper analysis by incorporating AI tools into feedback and review stages. Students may use AI to evaluate their drafts, identify gaps, and receive formative input before submitting final work. When students are asked to include both AI feedback and their revised drafts, instructors can assess improvement and thinking strategies. AI can assist instructors as well, offering preliminary evaluations that highlight patterns and reduce fatigue. These practices support metacognition, increase grading efficiency, and maintain academic rigor.

The chapter highlights how AI can enhance study habits through customizable review tools, such as flashcards, quizzes, and thematic breakdowns. Students can convert lecture content into active learning experiences, reinforcing retrieval practice and long-term comprehension. These strategies are especially helpful for students with learning differences, as they promote personalized and multimodal engagement. Teachers can also use AI to gamify learning, creating interactive formats that foster participation. Incorporating AI into study design ensures that students use the technology for reinforcement, not replacement.

References:

1. Johnson, D. W., & Johnson, R. T. (2020). *The role of self-assessment in education: A meta-analysis. Journal of Educational Psychology*, *112*(3), 453–472.
2. Smith, A., & Brown, P. (2021). *Interactive study tools and their impact on learning outcomes for students with learning disabilities. Learning Disability Quarterly*, *44*(2), 89–97.

3. Davis, C., & Lee, M. (2022). *Technology-enhanced critical thinking in higher education. Journal of Educational Technology, 38*(1), 15–29.
4. Thompson, G., & Harris, J. (2023). *Digital literacy in modern education: Ethical and practical approaches. Educational Review, 75*(4), 467–482.

Chapter 7: Saving Educators Time with AI

Educators constantly seek innovative methods to manage their growing workloads without compromising the quality of education. Artificial intelligence (AI) provides practical solutions for streamlining tasks, from grading and planning to creating interactive and engaging lessons. By leveraging AI tools, educators can enhance efficiency and devote more time to meaningful student interactions. These tools cater to all educational levels, ensuring adaptability and accessibility in diverse learning environments.

7.1 Generating Questions from Instructional Materials

Creating practical questions that align with instructional content can be time-consuming. AI simplifies this process by analyzing uploaded materials, such as textbooks, PowerPoints, or lecture notes, and generating questions tailored to students' comprehension levels. These questions help educators assess students' understanding while saving significant preparation time. Whether for elementary vocabulary lessons or advanced high school science topics, AI can adapt its outputs to suit the developmental needs of learners. Educators can guide the AI to produce questions that align with specific learning objectives, encouraging deeper engagement with the material.

Example for High School:

Prompt:

"Create five questions based on the uploaded biology chapter on photosynthesis that assess understanding and application of concepts for high school students."

AI Response:

1. Define the process of photosynthesis and its significance for plants.
2. What role does chlorophyll play in capturing sunlight?
3. Predict the effects of removing carbon dioxide from a plant's environment.
4. Explain the difference between light-dependent and light-independent reactions.
5. How do temperature and light intensity affect the rate of photosynthesis?

7.2 Grading Papers for Consistency and Fairness

Grading is an essential but time-intensive task that requires objectivity and consistency. AI can assist educators by evaluating student submissions against detailed rubrics, providing structured feedback, and highlighting areas for improvement. This ensures fair grading across all submissions and offers students actionable insights into their performance. Whether grading creative writing for elementary students or research essays for secondary students, AI enhances efficiency while maintaining evaluation standards. Additionally, educators can customize rubrics to reflect the unique goals of each assignment, allowing for greater alignment between teaching and assessment.

Example for Elementary Level:

Prompt:

"Grade the uploaded short story assignment for a fourth grader using the rubric provided. Provide feedback on creativity, sentence structure, and spelling."

AI Response:

- **Creativity (4/5):** The story shows imagination and originality, though adding more descriptive details could enhance engagement.
- **Sentence Structure (3/5):** Sentences are generally clear but lack variety in length and complexity.
- **Spelling (5/5):** No spelling errors were found, demonstrating careful proofreading.

7.3 Customized AI Tools for Specific Educational Needs

Custom AI tools are tailored to address the unique needs of educators and students at various educational levels. These tools can automate lesson planning, curricular mapping, and personalized feedback. For elementary educators, custom bots might develop differentiated reading assignments or interactive math games. Secondary educators might use these tools to generate case studies or simulate debates. In higher education, custom AI bots can assist with complex tasks, such as designing clinical simulations or mapping competencies to accreditation standards. These solutions enhance both teaching efficiency and student engagement.

Example for Secondary Education:

Prompt:

"Design a debate framework for an undergraduate public policy seminar on the regulation of emerging technologies. Include assigned roles, structured debate prompts, and recommended academic sources for preparation."

AI Response:

- **Roles:**
  Corporate technology executive, data privacy advocate, federal policymaker, ethics scholar, and consumer protection lawyer.
- **Debate Prompts:**
  1. Should the federal government require algorithmic transparency in consumer-facing AI applications (e.g., hiring platforms, healthcare triage tools)?
  2. What role should ethical considerations play in drafting national policy for AI surveillance and data collection?
- **Suggested Research Materials:**
  Peer-reviewed articles on AI bias and regulation, Federal Trade Commission (FTC) briefings on consumer data protection, academic analyses of the European Union's AI Act and General Data Protection Regulation (GDPR), and position papers from leading ethics scholars in technology.

7.3 Gamifying Education for Engagement

Gamification transforms traditional lessons into dynamic and interactive experiences, fostering student motivation and teamwork. AI facilitates gamification by generating themes, tasks, and progress-tracking systems. These tools, from elementary math games to high school historical simulations, make learning more immersive and enjoyable. Educators can design games aligned with curricular goals, using AI to streamline creation and ensure consistency. Gamification supports academic growth and develops soft skills like collaboration and critical thinking. The following are examples of using AI for gamifications:

Example for Elementary Level:

Prompt:

"Create a Jeopardy-style game using the uploaded multiplication problems. Include five categories and questions suitable for third graders."

AI Response:

Multiplication Jeopardy (3rd Grade Edition)

- **Categories:**
    1. Easy Facts (1–5)
    2. Word Problems

Category 1: Easy Facts (1–5)

Points	Question	Answer
100	What is 2 × 3?	6
200	What is 4 × 2?	8
300	What is 5 × 5?	25
400	What is 3 × 4?	12
500	What is 1 × 5?	5

Category 2: Word Problems

Points	Question	Answer
100	Emma has 3 bags with 2 apples each. How many apples in total?	6
200	A spider has 8 legs. How many legs do 2 spiders have?	16
300	There are 4 rows of chairs with 5 chairs in each row. How many total?	20
400	A pack of markers has 6 markers. If you buy 3 packs, how many markers?	18
500	A bakery bakes 7 loaves of bread every day. How many in 5 days?	35

This game encourages active learning by turning practice into play. Students review multiplication facts and apply math reasoning through age-appropriate word problems. Teachers can use the Jeopardy format to foster teamwork and provide real-time feedback during review sessions, helping reinforce fluency and problem-solving in a low-pressure environment.

Example for High School Level:

Prompt:

"Create a team-based role-playing game for a 10th-grade environmental science class focused on sustainability challenges. Include character roles, objectives, and point systems."

AI Response:

RPG Title: *Sustain the World*

Game Overview:

Students are divided into teams of 4–5, each representing a fictional community navigating global sustainability issues. Teams must manage resources, develop environmental policies, and build sustainable infrastructure—balancing ecological, economic, and social needs.

Game Setting:

Teams operate in regions inspired by real-world challenges including climate change, water scarcity, biodiversity loss, and energy shortages.

Character Roles (1 per student per team):

1. **Environmental Scientist** – Monitors pollution, biodiversity, and ecological impact.
2. **Policy Advisor** – Drafts environmental regulations and negotiates agreements with other teams.
3. **Economic Strategist** – Manages budget, proposes green technologies, and evaluates financial trade-offs.
4. **Public Relations Officer** – Communicates team decisions, runs public awareness campaigns, and tracks public support.
5. **Engineer/Technologist** *(optional)* – Designs sustainable infrastructure such as clean water systems and renewable energy sources.

Game Objectives:

Each team must:

- Maintain ecosystem health
- Reduce carbon emissions by 40%
- Ensure clean water access for 95% of citizens
- Achieve 80% renewable energy usage
- Avoid bankruptcy and maintain economic stability

Gameplay Structure (4 rounds = 4 years):

1. **Briefing:** A news update introduces a new global challenge (e.g., drought, oil crisis).
2. **Strategy Session:** Teams choose actions such as building infrastructure, creating policy, or negotiating with others.
3. **Action Submission:** Teams submit decisions to a game moderator (teacher or AI tool).
4. **Results & Negotiation:** Outcomes are shared. Teams may adjust strategy, collaborate, or compete for resources.

Point System:

Category	Max Points per Round	Criteria
Environmental Impact	20	Emissions, biodiversity, and water quality outcomes
Economic Stability	20	Budgeting, innovation, and sustainability of investments
Policy and Diplomacy	20	Lawmaking and inter-team negotiations
Public Awareness	20	Campaign effectiveness and clarity
Team Collaboration	20	Shared participation and cohesive decision-making
Total per Round	**100 points**	

Category	Max Points per Round	Criteria
Max Game Score	**400 points**	

This game encourages systems thinking, real-world problem solving, and collaboration. Students experience the complexity of sustainability decisions and build communication, negotiation, and leadership skills through an immersive learning environment.

Example for Secondary Education:

Prompt:

"Design a strategy-based classroom simulation for a business management course focused on entrepreneurship. Include rules, team roles, and scoring criteria. Use a Star Wars theme and talk like Yoda."

AI Response:

Simulation Title: “Galactic Start-Up Syndicate”

Younglings in the ways of business, train they must. Compete they will, as founders of startups across the galaxy. Survive market crashes, Imperial audits, and Hutt rivalries they must, if profit and peace they seek.

Simulation Setup

Split your padawans into teams of 4–5. Found a galactic venture, each team must. Choose wisely, their sector:

• Droid Manufacturing

• Exotic Foods from Naboo

• Lightsaber Craftsmanship

• Hyperdrive Couriers

• Holo-Entertainment Streaming

Yes… even Moisture Farming, if brave they are.

Roles Within the Order (Team Roles, hmmm)

1. **Jedi CEO (Chief Executive Officer)**
    - Align strategy with the Force (and market trends).
    - Make final calls in time of conflict.

2. **Holo-Credit Master (Chief Financial Officer – CFO)**
    - Track the credits, balance the ledgers, fund the future.
3. **Holonet Whisperer (Chief Marketing Officer – CMO)**
    - Control the galactic message. Design campaigns. Influence minds, they will.
4. **Systems Overseer (Chief Operating Officer – COO)**
    - Production and logistics their domain is. Supply lines and operations, they must master.
5. **(Optional for 5-player teams) Techno-Wizard (Chief Technology Officer – CTO)**
    - Develop gadgets, upgrade tech, automate the future.

Simulation Rules (Guided by the Force, they are)

1. Begin with 100,000 galactic credits. Use wisely, or vanish they will.
2. Each round, a galactic event strikes. A new market opens. A blockade forms. The Empire taxes.
3. Teams submit strategy reports: product decisions, marketing, tech upgrades, and expansions.
4. Use negotiation, teams can. Trade routes, customers, or secrets.
5. Final Jedi Council vote each round affects bonuses. Wisdom, not just credits, they must show.

Rounds of Play (4 There Are)

Each round = 1 galactic cycle (1 month in Earth time, or 1 class period in human school).

• Round 1: Launch & Branding (Establish product, market, and message)

• Round 2: Expansion & Innovation (Develop tech, open new systems)

• Round 3: Conflict & Crisis (Respond to surprise event: blockade, raid, scandal)

• Round 4: Legacy Round (Plan for the future, present to the Jedi Council)

🧮 Scoring Criteria (Points Earned, Not Just Credits)

Scoring Category	Max Points	What the Jedi Council Seeks
Profit & Financial Growth	25	Return on investment, balance sheets strong must be
Innovation & Design	20	Unique products, new systems, Force-infused creativity
Team Strategy & Leadership	20	Unity, logic, and delegation—like Jedi, not Sith
Marketing Mastery	15	Viral campaigns across the Holonet, influence powerful it is
Ethical Decisions	10	Bribery, pollution, sabotage? Dark side, that is. Choose integrity, they must.
Presentation to the Council	10	Final report to Jedi Council—clear, wise, and passionate it must be
Maximum Total: 100 points per cycle × 4 cycles = 400 points possible		
Bonus Points from Jedi Council: Up to 20		

🏆 Awards from the Council

• *Startup Jedi Masters* – Highest total score

• *HoloNet Heroes* – Most brilliant marketing campaign

• *Innovator Knights* – Most creative product or service

• *Balance of the Force* – Most ethical and balanced approach

🧠 **Want AI Integration, Do You?**

Use GPT as:

- A Jedi customer giving feedback.
- An unpredictable Sith investor offering risky credits.
- A Holonet feed to simulate social response.

7.4 Summary

Artificial intelligence offers educators meaningful ways to enhance teaching and learning across all educational levels. When used intentionally, AI can streamline routine tasks, support instructional planning, and increase student engagement without replacing human judgment. Throughout this chapter, examples demonstrated how AI can generate practice questions aligned with learning objectives, assist with consistent grading using rubrics, and offer real-time feedback to students. These tools reduce preparation time while improving instructional alignment and transparency.

Educators can also leverage AI to create personalized learning experiences. From developing differentiated reading materials in elementary settings to building immersive, theme-based simulations for high school and graduate students, AI can help structure lessons that meet diverse learner needs. Gamification, in particular, shows how AI can generate interactive, goal-oriented activities that build teamwork, critical thinking, and content mastery. Whether the goal is to simplify grading or to build a complex classroom simulation, AI can support educators by expanding their instructional toolkit while preserving student ownership of the learning process.

References

1. Brown, A., Thompson, R., & Lee, P. (2021). *Interactive learning strategies for students with dyslexia. Journal of Learning Disabilities*, *44*(2), 123–134.
2. Smith, J., Davis, C., & Robinson, M. (2022). *AI-assisted grading: Enhancing consistency in higher education. Journal of Educational Technology*, *39*(1), 56–72.
3. Carter, E., & Wilson, T. (2023). *Custom AI solutions for education: Transforming administrative and instructional practices. Educational Innovations Quarterly*, *48*(3), 234–251.
4. Davis, L., & Johnson, K. (2022). *Gamification in K-12 classrooms: Engaging students through interactive learning. Educational Psychology Review*, *34*(1), 45–67.
5. Baker, M., & White, R. (2020). *The impact of AI on teacher workload: A case study analysis. Journal of Educational Technology Research*, *32*(4), 345–359.
6. Taylor, H., & Green, P. (2023). *Developing adaptive AI tools for inclusive education. Journal of Special Education Technology*, *40*(2), 101–119.
7. Thompson, S., & Carter, A. (2021). *Leveraging technology for formative assessment: Trends and best practices. Journal of Assessment and Evaluation*, *37*(2), 89–105.
8. Wilson, G., & Turner, J. (2022). *The future of gamification: How AI is revolutionizing classroom experiences. Computers in Human Behavior*, *128*, 107134.
9. Li, R., & Zhang, Y. (2021). *AI in education: A review of emerging applications and challenges. Educational Technology Review*, *30*(3), 201–218.

Chapter 8: The Future of AI in Education: Learning Disabilities and AI

Artificial intelligence (AI) is revolutionizing education by offering innovative solutions to support students with learning disabilities. These tools enhance accessibility, deliver personalized instruction, and adapt to diverse learning needs. For educators, AI provides opportunities to create inclusive classrooms where all students can thrive. AI applications in special education include adaptive learning systems, text-to-speech tools, and interactive social robots. Exploring how these tools are used and their potential for future development can guide educators in maximizing their benefits.

8.1 How AI Supports Students with Learning Disabilities

AI supports students with learning disabilities by tailoring educational experiences to individual needs. Adaptive learning systems analyze performance data and adjust lesson difficulty accordingly. This approach ensures students progress at a pace suited to their abilities, addressing gaps in knowledge. For example, students with dyscalculia might receive customized math exercises focused on foundational skills before progressing to more complex topics. AI creates a more supportive learning environment by reducing frustration and increasing engagement.

Speech recognition tools are another example of AI's impact on education. Students with motor impairments or writing challenges can use voice commands to complete assignments or communicate ideas. These tools enable active participation in education while reducing physical barriers. Predictive text tools further enhance this experience by offering suggestions to simplify writing. Research shows that such interventions improve academic outcomes for students with learning disabilities by fostering independence and confidence (Smith et al., 2021).

Example Prompt:

"Provide personalized math exercises for a student struggling with division. Adapt the difficulty based on performance."

AI Response:

- Problem 1: What is 10 ÷ 2?
- Problem 2: Divide 25 by 5.
- If correct: Increase difficulty (e.g., What is 144 ÷ 12?).
- If incorrect: Provide a hint, such as "Think about how many times 5 fits into 25."

8.2 Personalized Learning Tools and Accessibility

AI-powered personalized learning tools are transforming accessibility in education by addressing students' unique needs. Text-to-speech tools allow students with dyslexia to convert text into audio, making reading materials more accessible. These tools also enhance comprehension, enabling students to engage with content they might otherwise find challenging. Similarly, visual schedules created by AI assist students with autism spectrum disorder (ASD) in managing transitions and routines, providing a structured framework for their day.

AI also helps remove language barriers for students with hearing impairments or who are nonverbal. Real-time captioning software ensures students can fully participate in classroom activities, while sign language translation tools enhance communication between students and teachers. These advancements promote equity by ensuring that all students have equal educational opportunities. Educators can use AI to customize these tools for specific classroom needs, fostering an inclusive learning environment that supports diverse learners.

Example Prompt:

"Create a visual schedule for a student with ASD, including visuals for a typical school day."

AI Response:

- 8:00 AM: Morning Meeting (Visual: Group of students in a circle).
- 8:30 AM: Reading Time (Visual: Open book).
- 9:15 AM: Snack Break (Visual: Apple and juice box).
- 9:30 AM: Math Lesson (Visual: Numbers on a chalkboard).

8.3 Case Studies: AI in Special Education

Real-world applications of AI in special education demonstrate its transformative potential. In one case, a reading assistant tool was implemented for students with dyslexia. The AI program assessed reading fluency, provided phonics exercises, and tracked student progress. Over six months, participating students showed measurable improvements in reading comprehension and self-confidence. Teachers noted that the tool reduced the time spent on one-on-one interventions, allowing them to focus on other instructional needs. Another example involved using social robots in classrooms for students with ASD. The robots modeled appropriate social behaviors, such as turn-taking and greeting peers, and engaged students in interactive activities. Educators observed increased participation and reduced anxiety among students during group interactions. Research suggests that AI-powered social robots create a safe and engaging environment for practicing social skills, improving real-world outcomes (Jones et al., 2022).

Example Prompt:

"Design an interactive activity for a social robot to teach turn-taking to a student with ASD."

AI Response:

- Activity: "Pass the Object."
    1. The robot explains the rules: "Take the ball when it is passed to you, say a word, and pass it to the next person."
    2. The robot models the behavior and encourages the student to participate.
    3. The robot offers praise and prompts as needed to reinforce positive behavior.

8.4 Areas for Future Research in AI and Special Education

While current AI applications in special education are promising, significant areas remain for further research. One priority is developing emotionally intelligent AI tools that can detect and respond to students' emotional states. For instance, an AI system could adjust tasks based on signs of frustration or disengagement, maintaining a supportive learning environment. Research into culturally inclusive AI tools is also essential, ensuring that tools address diverse linguistic and cultural needs.

Longitudinal studies are necessary to evaluate the long-term impact of AI on academic and social outcomes for students with learning disabilities. These studies could explore how sustained use of AI tools affects confidence, independence, and career readiness. Additionally, ethical considerations such as data privacy and algorithmic bias must be addressed to ensure equitable access to AI benefits. Collaboration between educators, technologists, and policymakers will be critical in advancing these research priorities.

8.5 Summary

AI is transforming special education by providing personalized support and increasing accessibility for students with learning disabilities. Tools such as adaptive learning systems, text-to-speech programs, and social robots empower educators to create inclusive and effective learning environments. Case studies highlight the practical benefits of AI, while future research offers opportunities to enhance its impact further. Educators can integrate AI thoughtfully to ensure that all students have the necessary resources to succeed.

References

1. Smith, J., & Davis, K. (2021). *The impact of adaptive learning systems on students with learning disabilities. Journal of Special Education Technology*, *39*(2), 101–117.
2. Jones, L., & Brown, M. (2022). *Social robots as peer models for children with autism spectrum disorder. Educational Robotics Review*, *25*(1), 34–48.
3. Carter, E., & Wilson, T. (2023). *AI tools for accessibility: Innovations in special education. International Journal of Educational Technology*, *49*(3), 345–362.
4. Taylor, H., & Green, P. (2022). *AI-driven tools for inclusive learning: Addressing diverse needs in education. Journal of Inclusive Education*, *40*(1), 67–85.
5. Li, R., & Zhang, Y. (2021). *Future trends in AI and special education. Educational Research Quarterly*, *30*(4), 289–305.

Appendix A: AI-Generated vs. Student-Tailored Prompts

Quick-Reference for Designing AI-Resistant Assignments

Prompt Type	Example Prompt	AI Response Quality	Educator Notes
Generic Prompt	Discuss the importance of teamwork.	High – AI can generate a full, polished response.	Avoid vague tasks. Students can bypass deep thinking with copy-paste responses.
Student-Tailored Prompt	Describe a group project you completed this semester. Explain two teamwork strategies your group used and one communication challenge you faced.	Low – Requires lived experience and course content.	Forces personal context and reflection. Difficult for AI to fabricate.
Generic Research Prompt	Compare local and global climate change policies.	Medium to High – AI can generate broad comparisons.	Risk of shallow content. Lacks personal or local connection.
Student-Tailored Prompt	Compare your city's climate policies to international agreements discussed in class. Include local data or legislation.	Low – Needs student-driven specificity.	Encourages real-world relevance and course-based application.

Generic Reflective Prompt	Write about a personal challenge in school.	Medium – AI can simulate basic emotional tone.	AI may mimic emotion, but responses often lack genuine depth or narrative detail.
Student-Tailored Prompt	Describe a time during this course when you struggled with an assignment. What steps did you take, and how did peer or instructor feedback influence your outcome?	Low – Requires detailed, course-specific input.	Encourages true reflection and self-awareness. AI cannot generate credible details.

☑ ***Design Tips for Educators:***

1. Require references to classroom discussions, lecture content, or peer feedback.
2. Ask for local, personal, or course-specific examples.
3. Use bracket-style prompts (e.g., "Create a summary of {event} based on {local news source}") to force input specificity.
4. Consider having students submit a screenshot of any AI-generated draft and explain how they revised it.

About the Author

Dr. Jeffery Adams, Ed.D., M.S., NRP, is an educator, innovator, and paramedic whose work sits at the crossroads of critical care, simulation-based education, and applied technology. With over 18 years in healthcare education, he has held leadership roles at institutions including The Ohio State University, Wright State University, Duquesne University, and Philadelphia College of Osteopathic Medicine. He currently serves as the Director of College of Health Sciences Simulation Center for Lipscomb University, where he leads efforts to integrate adult learning methodologies and simulation into clinical training across disciplines.

Dr. Adams' interest in education stems not from early academic success—but from struggle. Labeled with a learning disability in grade school, he later realized it was not a disability but a different way of learning. It was through andragogy, technology, and hands-on experience that his learning flourished. That journey helped shape his belief that education should spark curiosity, promote autonomy, and embrace difference.

As a lifelong paramedic, he remains grounded in the realities of patient care—an experience he credits with keeping him humble and connected. As a simulation designer, he thrives on crafting experiences that push learners to think, adapt, and connect theory to action. As a medical simulation and AI consultant he has built AI-driven tools to support curriculum development and faculty workflow, not to replace teaching but to enhance creativity.

Outside the classroom, Jeffery spent over 20 years as a martial arts instructor, teaching Shaolin Kung Fu and Tai Chi. He often describes martial arts as the most honest form of learning: "You forget—you get hit." For him, the process of analyzing the strategy behind a form or movement echoes the work of educational design—understanding the why, not just the how. Tai Chi, in particular, taught him how to stay calm under pressure—whether in a simulation lab, a classroom, or the back of an ambulance.

Jeffery is the author of multiple publications on medical simulation and is a frequent presenter at international conferences, including IMSH. *Engaging Minds* reflects his belief that AI should not be feared—it should be used to encourage critical thought, not shortcut it. "I like to throw out ideas," he says, "because you never know who's going to take a small one and run with it."

For collaborations, speaking engagements, or further conversation, Dr. Adams can be reached at:

- Email: jefferyad@icloud.com
- LinkedIn: linkedin.com/in/jeff-adams-a1b306316

Made in the USA
Monee, IL
04 November 2026

67383561R00066